7 Chemistry of Plant Protection

Herbicide Resistance - Brassinosteroids, Gibberellins, Plant Growth Regulators

With contributions by
G. Adam, S. O. Duke, D. Gross,
M. Lischewski, V. Marquardt,
K. C. Vaughn, B. Voigt

With 10 Figures and 4 Tables

Springer-Verlag
Berlin Heidelberg New York
London Paris Tokyo
Hong Kong Barcelona
Budapest

Editor-in-Chief

Prof. Dr. *W. Ebing*
Biologische Bundesanstalt für Land- und Forstwirtschaft
Königin-Luise-Str. 19
1000 Berlin 33/FRG

This series continues the handbook
"Chemie der Pflanzenschutz- und Schädlingsbekämpfungsmittel"
edited by R. Wegler

ISBN-13: 978-3-642-48789-7 e-ISBN-13: 978-3-642-48787-3
DOI: 10.1007/ 978-3-642-48787-3

Library of Congress Cataloging-in-Publication Data.
Herbicide resistance-brassinosteroids, gibberellins, plant growth regulators / with contributions by G. Adam ... [et al.].
p. cm. — (Chemistry of plant protection; 7) Includes index.
1. Herbicide resistance — Physiological aspects. 2. Gibberellins. 3. Brassinolide. 4. Plant regulators. I. Adam, G. (Gerold), 1933–. II. Series. SB951.4.H445 1991
632'.954–dc20 91-27616CIP

Softcover reprint of the hardcover 1st edition 1991

Typesetting: Th. Müntzer, Bad Langensalza; Printing: Heenemann, Berlin;
Bookbinding: Lüderitz & Bauer, Berlin
52/3020-543210 — Printed on acid-free paper

Managing Editors

Editorial Board

Prof. J. Miyamoto
Laboratory of Biochemistry
and Toxicology
Takarazuka Research Center
Sumimoto Chemical Co. Ltd.
2-1, 4-Chome Takatsukasa
Takarazuka-Shi
Hyogo-Ken 665, Japan

Dr. John V. Schloss
Research Supervisor
Central Research and Development
E. I. DuPont de Nemours
Wilmington, DE 19898, USA

Dr. L. G. M. Th. Tuinstra
State Institute of Quality Control
of Agricultural Products
Bornesteeg 45
NL-6708 PD Wageningen

Prof. Dr. Andrzej Zabża
Technical University of Wroclaw
Institute of Organic and Physical Chemistry
Wybrzeze Wyspianskiego 27
PL-50-370 Wroclaw

Honorary Member
Prof. Dr. Richard Wegler
Auf dem Forst
D-5090 Leverkusen-Schlebusch

Editorial

In 1986 Springer-Verlag started to published "chemistry of Plant Protection" as the successor to the well established handbook "Chemie der Pflanzenschutz- und Schädlingsbekämpfungsmittel", edited by R. Wegler. The series "Chemistry of Plant Protection" publishes critical review articles on new aspects and developments in the field of plant protection and pest control, ecotoxicological behaviour and properties of pesticides including biological and biochemical processes, residue analysis as well as toxicity testing. The contributions are written in English by invited authors; all manuscripts are reviewed. The aim of the publisher and the board of editors is to produce high quality review on the rapidly developing field of plant protection chemistry. Among others topics to be covered are: low dosage compounds, target sites and mode of action, formulations, environmental fate and compatibility, side effects, selectivity and efficacy, biosynthesis inhibitors or screening of natural products with pesticidal activities.

From Volume 9 onwards each volume will very strongly be oriented towards a volume topic.

In the four year period of 1986–1990, six volumes appeared. Beginning with Volume 7 a new Managing Editorial Board and a new Editorial Board were established. I took over the responsibility as Editor-in-Chief of the series and am happy to cooperate with an active staff of Managing Editors and an International Editorial Board.

The contributions in Chemistry of Plant Protection are written to a fairly advanced level and should be of interest to graduate students, research fellows and practicing scientists. I am sure that the subject matter treated is of high interest to scientists in industry, universities, as well as in governmental and regulatory bodies. Also natural product chemists, biologists, and organic chemists will benefit from the contributions.

It is with great satisfaction that I thank all authors and editors involved for the time they devoted to this effort.

Berlin, June 1991 W. Ebing

Table of Contents

Plant Growth Regulatory Substances Both of Microbial and Plant Origin

D. Gross

Institute of Plant Biochemistry, O-4050 Halle (Saale), Weinberg, FRG

Natural products of microbial and plant origin which exhibit plant growth regulating activities are reviewed. Phytohormones such as auxins, cytokinins, gibberellins, abscisic acid and ethylene are not included. The review only surveys low molecular weight compounds that have been chemically characterized and have been reported to induce regulatory responses in plants. The chemical structures, the occurrence and physiological activities of these endogenous plant growth regulators are represented.

The article deals mainly with natural plant growth regulatory substances either isolated and structurally elucidated later than 1980, or which have been under further investigation during the last few years. The classification of the plant growth regulators, presented here, is based on their chemical structure. Details of the methods and techniques used for isolation, chemical identification and bioassay are not given.

Chemistry of Plant Protection 7

1 Introduction

A characteristic feature of microorganisms and plants is their capacity to synthesize a high number of organic compounds which show an enormous chemical diversity. These natural products are designated as so-called microbial, secondary metabolites and secondary plant products. For many years it has been debated whether these metabolites are only waste products or whether they exert a specific, physiological or biochemical action on a target organ. Nevertheless, numerous biologically active compounds have been demonstrated which act within the organism producing these metabolites or, which exhibit different biological activities when used in a bioassay.

This review includes natural products of microbial and plant origin which possess plant growth regulatory properties. The well-known plant hormones, or phytohormones, such as auxins, cytokinins, gibberellins, abscisic acid, and ethylene, are broadly distributed within the plant kingdom and are, in fact, present in all higher plants. They are biosynthesized by the plant and transported from the site of production to the site of action where they, in general, produce their specific activity at very low concentrations. Growth and developmental processes of all plants are regulated by a complex interaction of plant hormones. These hormones have been the subject of numerous reviews and monographs dealing with different aspects of phytohormone research. For recent reviews see Davis (1987) [1], MacMillan (1980) [2], Rivier and Crozier (1987) [3], Sembdner et al. (1987) [4], and Takahashi (1986) [5].

Since the recent progress on phytohormones is summarized yearly in Chapt. V of Progress in Botany (Dörffling and Crocoll (1988) [6]; Sembdner et al. (1989) [7]) these hormones are excluded from this review.

In addition to the phytohormones, plants also contain numerous other biologically active compounds which elicit either inhibitory or stimulatory responses in the intact plant or when supplied exogenously to a plant bioassay. Some of these responses are implicated in the control of plant growth and development.

In the following, native plant growth regulators other than the phytohormones are presented. These naturally occurring compounds may be either widely distributed within plant species or restricted to a few, and they may be produced in large amounts or in minute quantities. Several of them may play a significant role in the regulation of growth and developmental processes of the host plant in vivo. Additionally, the responses evoked by exogenous application are often observed in plant species unrelated to species from which the plant growth regulating compound was isolated. A few groups of these new plant constituents that influence plant growth may now be considered phytohormone-like compounds because of their biological activity and their broad distribution in the plant kingdom.

Natural products that exhibit growth-regulating activity have been isolated not only from plant tissues but also as secondary metabolites from microorganisms, predominantly fungi.

This paper also surveys fungally-derived natural products that affect growth of higher plants. Microbial metabolites with only phytotoxic activity are excluded.

However, the following review deals only with low molecular-weight natural products from plant and microbial sources which have been chemically characterized. Details of the methods and techniques used for isolation, structural elucidation and bioassays are not included.

Some previous reviews covering naturally occurring plant growth regulators have been published by Bearder [8], Gross [9, 10], Letham [11], Mandava [12], McLaren [13], Schreiber [14], Sembdner and Gross [15], and Takahashi [16]. Therefore, this article deals mainly with natural plant growth regulatory substances which have either been isolated and structurally elucidated later than 1980 or which have been under further investigation during the last few years. Consequently, this review complements and supplements previous review articles. For a survey of the older literature on native plant growth substances described before 1980, the reader is referred to the reviews mentioned above.

In the following overview, plant growth regulatory substances, both of microbial and plant origin, are generally reviewed with respect to their chemical structure. Based on their physiological activity on general or specific processes of plant growth and development some groups of endogenous plant growth regulating substances have received particular attention. Thus, they will be discussed in more detail. The majority of natural products exhibiting plant growth regulating properties, however, can be described only briefly.

2 Aliphatic Compounds and Biogenetically Derived Compounds

A number of aliphatic alcohols and carboxylic acids have long been known to inhibit the growth of several plants or excised plant organs (cf. [8, 10, 11, 12]). Moreover, some new results are presented.

2.1 Alcohols

The long-chain primary alcohol, 1-triacontanol **1**, has previously shown to be a natural constituent of several plants and of plant waxes. Many studies have shown that exogenously applied **1** can enhance the growth and yield of many important crop species, such as rice, barley, corn and others (cf. [8, 10, 27]). Because of its significant plant growth stimulating effects after exogenous application at concentrations up to 10^{-8} M, triacontanol has been intensively studied under different aspects in the past few years, for example, seed germination and early growth [17, 18, 19], growth and assimilation [20], ATPase activity [21], growth and development [22], photosynthesis and photorespiration of C_3- and C_4-plants [23], peroxidase, indolylacetic acid and plant growth [24], absorption and transport of Rb^+ and PO_4^{3-} ions [25] and the regulation of plant growth [26, 27, 28, 29]. These papers indicate the important role of triacontanol in plant growth regulation.

Some new routes for preparing this plant growth regulator synthetically have also been achieved [30, 31, 32].

Very recently triacontanol was shown to elicit rapidly a water-soluble second messenger in rice shoots after foliar application at nanomolar doses [33]. This messenger moves quickly throughout the plant. After treatment of rice seedling shoots with 1-octacosanol **2**, shown to be also a plant constituent, another second messenger is produced that inhibits the growth stimulating activity of triacontanol. Both alcohols probably act on plasma membrane to produce the second messengers. Triacontanol is assumed to be an endogenous plant hormone-like substance which regulates important metabolic functions of plants, including the enhancement of growth processes.

$$CH_3{-}(CH_2)_{28}{-}CH_2OH$$

1

$$CH_3{-}(CH_2)_{26}{-}CH_2OH$$

2

From avocado, *Persea americana*, seeds four non-auxinic rooting promotors were isolated and identified as 1,2,4-trihydroxy-*n*-heptadeca-16-ene **3**, 1,2,4-trihydroxy-*n*-heptadeca-16-yne **5** and their 1-O-acetyl derivatives **4** and **6** [34, 35, 36]. The rooting activity was verified by mung bean bioassay. The triol **5** showed the highest activity. Therefore it appears that compounds of this structural type

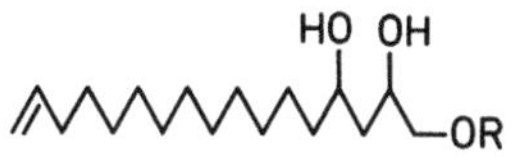

3 R = H
4 R = $COCH_3$

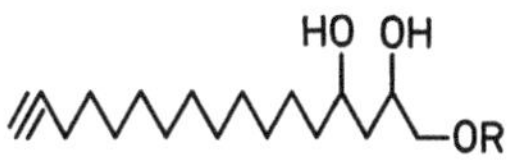

5 R = H
6 R = $COCH_3$

containing and acetylenic moiety are more active than those with an olefinic moiety. In 1971, compound **4** was isolated from avocado mesocarp and from some other plant tissues [37]. In contrast to the recent results, obtained by a mung bean bioassay system, compound **4** was shown to be a growth inhibitor of wheat coleoptiles and soybean callus, whereas compound **3** was found to be inactive. It seems necessary to examine these endogenous plant growth regulators by the same bioassays in order to precisely compare and evaluate their biological properties.

2.2 Carboxylic Acids

For many years fatty acids had been known to possess plant growth regulatory activity. Subsequent work has shown that unsaturated C_{18} fatty acids, such as oleic, linolic and α-linolenic acid from oat, *Avena sativa*, are active in senescence

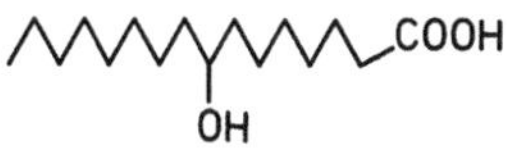

7

promotion both in light and darkness [38]. Recently, the plant growth inhibitory activity of certain fatty acids and their methyl esters as well as of octanol and mono-, di- and triglycerides was examined [39]. Strong inhibitory activity by the *Avena* coleoptile bioassay has been found to be limited to those fatty acids with chain length of 8 to 12 carbon atoms. This result is in accordance with earlier experiments [40].

Linolic acid and (+)-8-hydroxyhexadecanoic acid **7** from *Lygodium japonicum* were found as inhibitors of spore and pollen germination in *L. japonicum*, *Camellia japonica* and *C. sinensis* [41].

Recently, the methyl ester of 3-methyl-8-hydroxy-4-decenoic acid **8** was isolated as a metabolite of *Aspergillus niger* [42]. This novel fatty acid methyl ester inhibited the germination of cress and lettuce seeds.

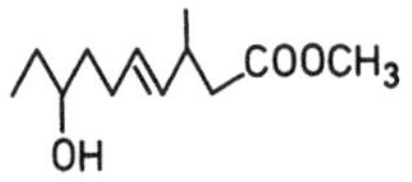

8

From surface lipids of *Nicotiana glutinosa* two new types of sucrose esters were isolated and identified as (2,3,4-tri-*O*-acyl-α-D-glucopyranosyl)-(3-*O*-acetyl)-*β*-D-fructofuranoside **9** and its 3-*O*-deacetyl derivative **10** [43]. The main fatty acid moieties were acyl residues of acetic, propionic, 2-methylbutyric, pentanoic, 4-methylpentanoic, hexanoic, 4-methylhexanoic, 5-methylhexanoic, heptanoic, and octanoic acids. Both these short-chain fatty acids and the corresponding sucrose esters inhibited tobacco seed germination and shoot growth of barnyard grass, *Echinochloa crus-galli*. Similar findings with fatty acid esters of glucose, sucrose and trehalose on *Avena* straight growth have previously been reported [44].

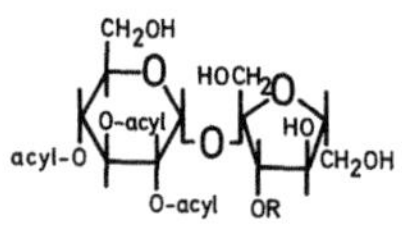

9 R = acetyl
10 R = H

Very recently, *N*-(3-hydroxyisotetradecanoyl)anatabine **11** and several acylnornicotines **12** were isolated from surface lipids of *Nicotiana repanda*, *N. stocktonii* and *N. nesophila* and structurally identified [45]. They have been found to inhibit tobacco seed germination and were compared with a number of synthetic analogues concerning their seed germination inhibitory activity.

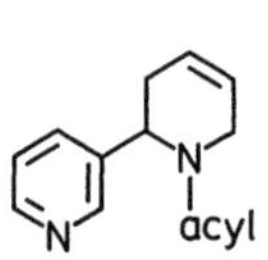

11

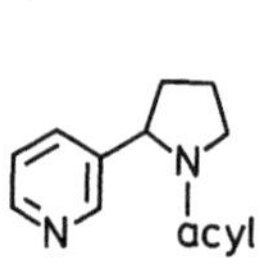

12

Moreover, some unusual di- and tricarboxylic acids were identified as endogenous plant growth regulators. Isopropylmalic acid **13** was shown to be a con-

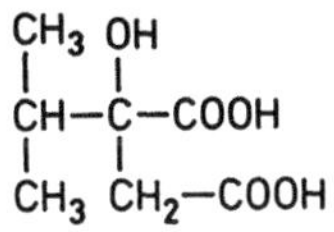

13

stituent of ripe fruit and seeds of *Sorbus aucuparia* and to act as an inhibitor of germination and growth [46]. (+)-2-Hexyl-3-methylenesuccinic acid (hexylitaconic acid) **14**, a metabolite produced from the fungus, *Aspergillus niger*, and excreted

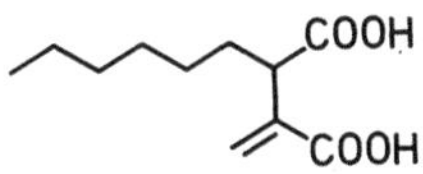

14

into the culture medium, was characterized as a root growth stimulating substance of lettuce and rice seedlings [47]. Extracts of oat, *Avena sativa*, husks contain some germination inhibitors [48]. One of them was characterized as *n*-pentane-1,3,4-tricarboxylic acid (dihydrohematimic acid) **15**.

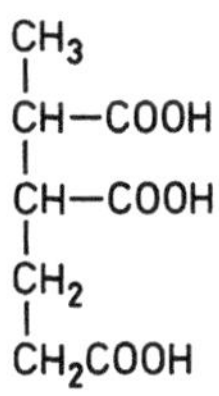

15

2.3 Cyclopentanoidal Fatty Acids

Prostaglandins, jasmonic acid and several structurally related C_{12}-compounds possessing a cyclopentanoidal structure are biogenetically derived from unsaturated fatty acids which are built up by the well-known pathway from acetyl-CoA.

Prostaglandins are known to be hormones of animals, particularly of mammals. Jasmonic acid and methyl jasmonate have been shown to be ubiquitous plant constituents which exhibit significant plant growth regulatory properties. Therefore, they may now be considered endogenous phytohormone-like substances that are discussed here in more detail.

2.3.1 Prostaglandins

The prostaglandins E_2 **16** and $F_{2\alpha}$ **17** have been found in the red alga, *Gracilaria lichenoides* [49]. Later $PGF_{2\alpha}$ was detected in *Kalanchoe blossfeldiana* [50]. Recently the prostaglandins F, E, A and B were shown to be constituents of *Populus*

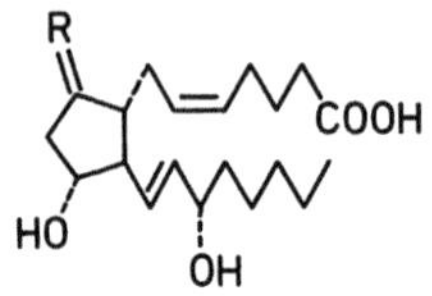

16 R = O
17 R = α-OH, H

balsamifera and *Larix sibirica* [51, 307]. It is not yet known, however, whether or not these compounds possess any biological activity within the plant. In contrast, jasmonic acid **18** and some of its naturally occurring derivatives are structurally related to the prostaglandins and have been found to possess significant senescence-promoting and growth inhibiting activities. As shown in the following, much new and interesting work has been reported on this new type of endogenous plant growth regulator.

2.3.2 Jasmonic Acid and Related Compounds

In 1962, (—)-methyl jasmonate **19** was isolated from *Jasminum grandiflorum* as an odoriferous compound and structurally elucidated [52]. About ten years later, jasmonic acid **18** was found as a plant growth inhibiting substance produced by the fungus *Botryodiplodia (Lasiodiplodia) theobromae* [53]. Compound **18** and 9,10-dihydrojasmonic acid **20**, both conjugated with isoleucine, have been

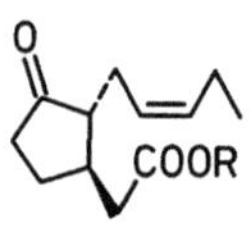

18 R = H
19 R = CH_3

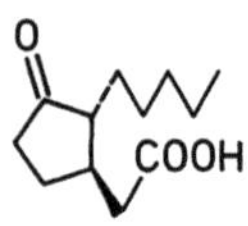

20

found to be metabolites produced by the fungus *Gibberella fujikuroi* [54]. In 1980, (—)-methyl jasmonate **19** was isolated from *Artemisia absinthium* and identified as a senescence-promoting substance [55]. About the same time, (—)-jasmonic acid **18** was isolated as an endogenous plant growth regulator from the pericarp of immature fruits of *Vicia faba* [56]. Later jasmonic acid and its methyl ester were shown to occur generally in plants [57, 58, 59, 60, 61].

Jasmonic acid and/or methyl jasmonate exhibited different, often ABA-like physiological activities, at least after exogenous application to plants, for instance senescence-promoting activity in excised oat leaves [55, 62, 63, 64], inhibition of cytokinin-induced soybean callus growth [65] and inhibitory effects on longitudinal growth of seedlings of rice, oat and wheat [66, 67, 68, 69, 70]. Further earlier findings on physiological effects are summarized by Miersch and Sembdner [69] and by Sembdner and Gross [15]. The role of jasmonate as hormonal regulator or stress factor in leaf senescence has been very recently discussed by Parthier [338].

Due to the different physiological activities and the ubiquitous occurrence of jasmonic acid and methyl jasmonate in higher plants both substances are the most important members of a new group of endogenous plant growth regulators, that have received much attention. During the last few years, therefore, these naturally occurring plant growth regulators have been studied with a particularly high intensity. Some new interesting results on their biosynthesis, metabolism and physiological role are described briefly.

In addition to jasmonic acid and methyl jasmonate some amino acid and glucose conjugates and several structurally related compounds have been isolated from different plant and fungal sources.

The phenylalanine methyl ester of 12-acetoxyjasmonic acid was found in *Praxelis clematidea* [71], whereas conjugates of (—)-jasmonic acid **18** with the amino acids tyrosine and tryptophan, both linked by a peptide bond to the carboxylic group of jasmonic acid, have been found to occur in *Vicia faba* [72, 73]. These conjugates were characterized as *N*-[(—)-jasmonoyl]-*S*-tyrosine and *N*-[(—)-jasmonoyl]-*S*-tryptophan, respectively. The biological activity of some jasmonic acid conjugates in a senescence and greening assay using barley leaf segments has also been reported [74]. However, it appears not yet clear whether these conjugates inhibit their biological effects *per se* or after enzymatic release of the biologically active jasmonic acid.

The most important analogue of jasmonic acid seems to be its natural stereoisomer (+)-7-*iso*-jasmonic acid **21** (7-*epi*-jasmonic acid) which was shown to be a metabolite of the fungus *Botryodiplodia theobromae* (synonym *Lasiodiplodia theobromae*) [69, 75]. This fungus does not produce (—)-jasmonic acid. The presence of **21** or its methyl ester **22** has also been demonstrated in higher plants, such as *Citrus limon* [76] and *Vicia faba* [78, 79].

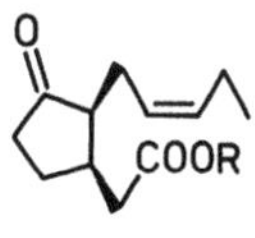

21 R = H
22 R = CH_3

A structurally related compound, (+)-cucurbic acid **23**, has been isolated from *Cucurbita pepo* where it is accompanied by its 6-*O*-glucoside and by methyl cucurbate glucoside [80, 81].

Moreover, jasmonic acid **18**, 6-*epi*-7-*iso*-cucurbic acid **24** and 4,5-didehydrojasmonic acid **25** have been detected in three species of *Equisetum* [82]. (—)-Jasmonic acid **18**, (+)-7-*iso*-jasmonic acid **21**, (—)-9,10-dihydrojasmonic acid **20**, 3,7-didehydrojasmonic acid **26**, and (+)-6-*epi*-7-*iso*-cucurbic acid **24** have been shown to occur in immature fruits of *Vicia faba* [79]. (+)-Cucurbic acid **23**, (+)-9,10-dihydro-7-*iso*-jasmonic acid **27** and (+)-11,12-didehydro-7-*iso*-jasmonic acid **28** were isolated from *Botryodiplodia theobromae* and structurally elucidated [75]. These naturally occurring jasmonic acid isomers or derivatives were shown to possess plant growth regulating activities, but at different levels and dependent on the bioassay system used.

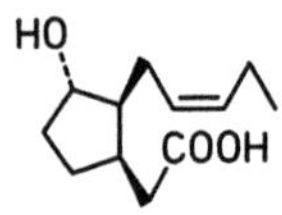

23

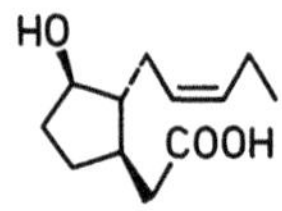

24

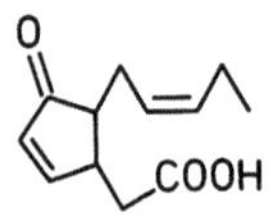

25

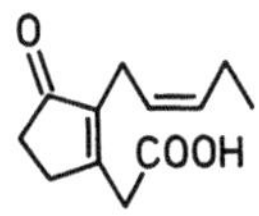

26

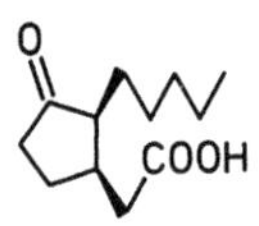

27

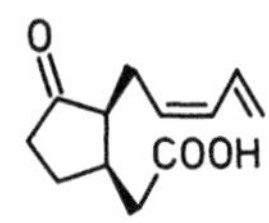

28

Many synthetic routes to methyl jasmonate have been reported (cf. [83] and references therein). The syntheses of racemic (2-^{14}C) jasmonic acid [84] and (2-^{14}C) 9,10-dihydrojasmonic acid [85] as well as of methyl 7-*iso*-jasmonate and methyl cucurbate [86] have also been achieved.

The biosynthesis of jasmonic acid and biogenetically related compounds in plants have been shown to follow a prostaglandin-like route starting from linolenic acid [87, 88, 89, 90]. The enzymes catalyzing the conversion to jasmonic acid have also been studied.

The major metabolites of (2-^{14}C) (±)-9,10-dihydrojasmonic acid supplied to barley seedlings were identified as (—)-9,10-dihydro-11-hydroxyjasmonic acid and its *O*(11)-*β*-D-glucopyranoside [91].

During the last few years the various physiological effects in plant growth regulation caused by exogenous application of jasmonic acid or methyl jasmonate to plants were extensively studied with different plant bioassay systems. Very recently jasmonic acid and the *O*-glucoside of 12-hydroxyjasmonic acid **161** were reported to occur in potato leaves as potent potato tuber-inducing substances [92, 93, 312, 339]. Methyl jasmonate has been shown to stimulate the ethylene production in tomatoes [94, 95, 96] and apple fruits [97, 98].

Recent interesting experiments were focused on the markedly promotion of leaf tissue senescence. It was found that treatment of barley primary leaf segments with methyl jasmonate led to the accumulation of *de novo*-synthesized polypeptides. In the course of which simultaneously a drastic degradation of ribulose-1,5-bisphosphate carboxylase and a decrease of chlorophyll have been found to occur [99, 100, 101]. Moreover, jasmonic acid and methyl jasmonate were able to induce and accumulate such specific abundant proteins in leaf tissues of a large number of plant species, both monocotyledons and dicotyledons [102]. Treatment of leaves or suspension cultures of soybean with jasmonic acid also increased the level of specific polypeptides [103]. However, the exact role of jasmonic acid in plant growth regulations and its mode of action in plant cells is still unknown and needs further investigation.

3 Alicyclic Compounds

Pevalic acid **29** is a metabolite produced from the fungus, *Penicillium valiabile*, that affect chlorophyll retention in oat leaf segments [104]. The cyclohexene derivative streptol **30**, isolated from the culture filtrate of an unidentified *Strepto-*

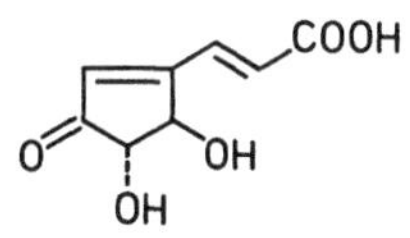

29

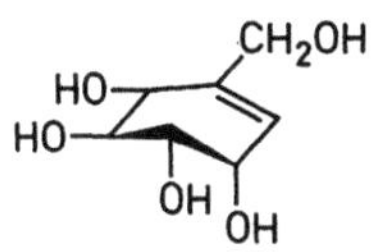

30

myces sp., was found to inhibit root growth of lettuce seedlings at concentrations of 13 ppm [105]. It belongs to the D-series of cyclitol. The saturated derivative of streptol was completely inactive. From this result it was concluded that the allylic alcohol system might be important for its biological activity.

4 Aromatic Compounds

The phenolics elaborated by plants and to a lower extent by microorganisms range in structural type over a wide spectrum from simple phenols and hydroxy benzoic acids to complex compounds such as tannins and humic acids. Phenolic compounds are of very wide occurrence in higher plants, many of them are ubiquitous. Often phenolics occur as *O*-glycosides.

A number of natural phenolic compounds like substituted benzoic and cinnamic acids, quinones, flavonoids, coumarins and other aromatic compounds have been found to be active inhibitors of some plant growth processes although they apparently do not show very specific effects in vivo (see [8, 9, 10, 11, 12]). Moreover, some new results on plant and microbial phenolic compounds that exhibit plant growth regulatory activity are presented.

4.1 Benzoic and Cinnamic Acid Derivatives and Related Phenolic Aldehydes and Ketones

Benzoic acid **31** and several of its hydroxylated derivatives, such as *p*-hydroxybenzoic acid **33**, gentisic acid **35** and vanillic acid **37** as well as *p*-hydroxybenzaldehyde **40** and vanillin **42**, all found in leaves of *Imperata cylindrica*, have been reported to be inhibitors of tomato radicle elongation [106]. Benzoic acid **31** and salicylic acid **32** have been found to occur in *Lemna paucicostata* and *L. gibba* and to induce effectively flowering in various strains of *Lemna paucicostata*, *L. gibba* and *L. minor* [107, 115, 290, 291], but the flower-inducing effect of **31** was strongly modified by plant hormones. Therefore, benzoic acid may interact with other factors to influence the flowering response in *Lemna*. Benzoic acid **31** and *p*-hydroxybenzoic acid **33** have been isolated from pollen of *Pinus densiflora* as inhibitors of germination and tube elongation of pollen from *Camellia japonica* and some other species [108, 109].

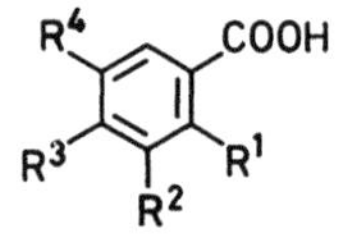

	R^1	R^2	R^3	R^4
31	H	H	H	H
32	OH	H	H	H
33	H	H	OH	H
34	H	OH	OH	H
35	OH	H	H	OH
36	H	OH	OH	OH
37	H	OCH_3	OH	H
38	H	OCH_3	OH	OCH_3

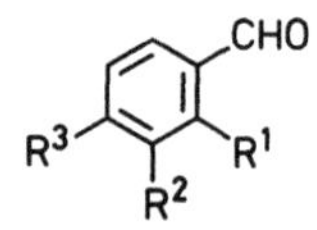

	R^1	R^2	R^3
39	OH	H	H
40	H	H	OH
41	H	OH	OH
42	H	OCH_3	OH

An agent that triggers pronounced heat production in flowers of some thermogenic plants is denoted as calorigen. Salicylic acid **32** from male flowers of *Sauromatum guttatum* has been shown, at least in some *Arum* lilies, to act as an endogenous regulator of heat production [110, 366]. The exogenous application of 0.13 μg salicylic acid/g fresh weight to appendix tissue of *S. guttatum* led to temperature increase of as much 12 °C. It is assumed that salicylic acid serves as a naturally occurring thermogenic trigger.

The fungus, *Rhizoctonia oryzae*, produces *p*-hydroxybenzoic acid **33** as sheath spot phytotoxin which inhibited root growth of lettuce seedlings [111].

From seed balls of red beet, *Beta vulgaris*, 8 phenolic inhibitors were isolated and identified as the hydroxylated benzoic acids **34**, **37** and **38**, *p*-hydroxybenzaldehyde **40**, protocatechualdehyde **41**, vanillin **42**, and syringaaldehyde [112]. Using lettuce seeds the phenolic aldehydes were found to be more potent germination inhibitors than phenolic acids. *p*-Hydroxybenzaldehyde **40** and anisalcohol are phenolic inhibitors from male flowers of *Cucurbita pepo* that show strong growth inhibition of lettuce seedlings [113].

Recently 4-*O*-methylorsellinic acid **43**, barbatic acid, evernic acid, diffractaic acid, *β*-orcinolcarboxylic acid **44**, 4-*O*-demethylbarbatic acid **45**, orsellinic acid, lecanoic acid and 3α-hydroxydiffractaic acid have been discovered in the lichen, *Usnea longissima*, and they exhibited growth-inhibition activity in lettuce seedlings [114]. *β*-Orcinol type depsides showed stronger activity than orcinol type depsides.

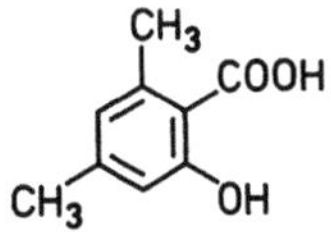

43

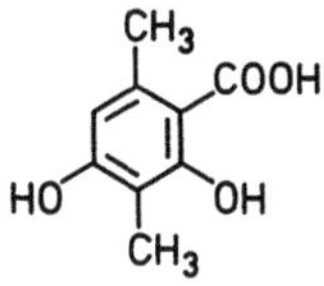

44

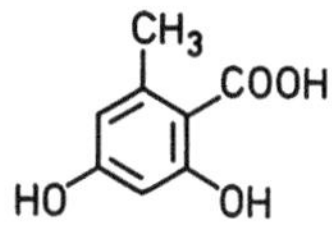

45

Phenylethanol **46**, phenylacetic acid **47**, phenoxyacetic acid **48** and *p*-methoxyphenylacetic acid **49** are produced by the fungus, *Aspergillus niger*, and found to inhibit germination [42]. Phenylacetic acid **47** which displays auxin-like activity has also been discovered in a range of crop plants like pea, wheat, barley, maize, sunflower, tomato, and tobacco [116]. 3,5-Dihydroxy-2-octanoylphenylacetic acid **47a** was isolated from a *Phoma* sp. and shown to be active in plant growth regulation [117].

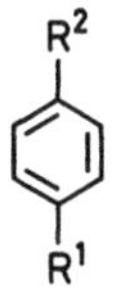

	R1	R2
46	H	CH_2CH_2OH
47	H	CH_2COOH
48	H	OCH_2COOH
49	OCH_3	CH_2COOH

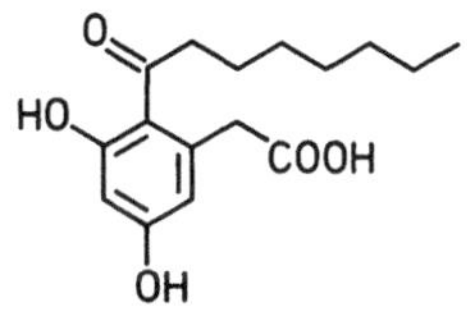

47a

A phenolic acid, dehydrochorismic acid **50**, isolated from *Pinus densiflora* pollen is an inhibitor of pollen tube elongation [118].

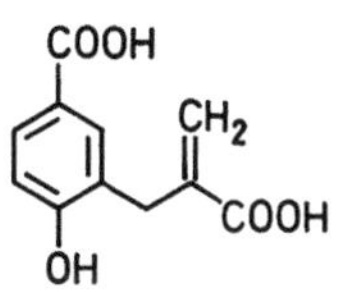

50

Phenylglyoxal **51** from *Pharbitis purpurea* and its artificial condensation product with acetone, 2-hydroxy-1-phenyl-1,4-pentanedione, exhibited flower-inducing activity in *Lemna paucicostata* [119]. 3,4-Dimethoxyacetophenone **52** isolated from oat husks is an inhibitor of germination and growth [48].

Grandinol **53** from the leaves of *Eucalyptus grandis* [120] and *E. pulverulenta* [121] is a phenolic ketone with strong germination inhibitory activities. Recently, a number of analogs have been synthesized and tested for germination inhibition of cress seeds. Some of the compounds structurally related to grandinol **53** were found to have strong inhibitory properties [122, 124]. The antigermination activity of grandinol and its derivatives required both the formyl and ketonic groups [123].

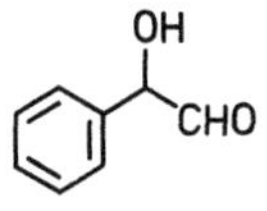

51

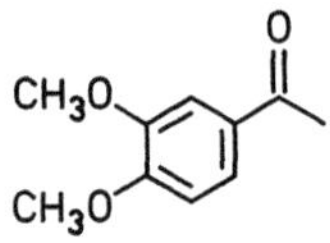

52

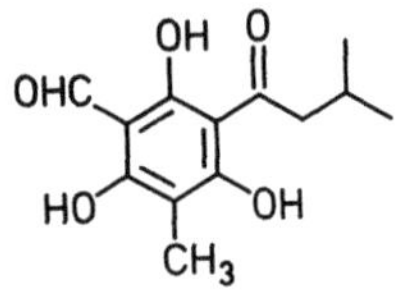

53

p-Coumaric acid **55** and its *o*-isomer found in leaves of *Imperata cylindrica* were shown to inhibit tomato radicle elongation [106]. Dihydroconiferyl alcohol **55a**, already known as an endogenous plant growth regulating compound was detected in *Acer pseudoplatanus* as a cell division factor [125]. It stimulated growth of soybean and tobacco callus and interacted synergistically with kinetin.

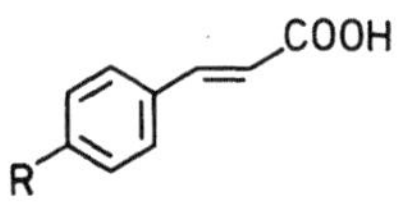

54 R = H
55 R = OH

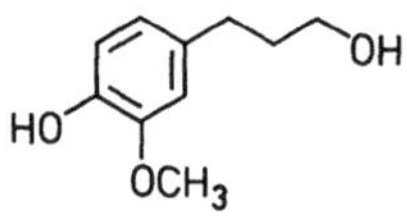

55a

Raphanusol A and B have been isolated from light-exposed seedlings of *Raphanus sativus* as new plant growth inhibitors, which inhibited the growth of etiolated radish and lettuce seedlings [126, 127, 128]. Raphanusol B was characterized as glucosyl ester of 3,5-dimethoxy-4-hydroxycinnamic acid **56**, whereas in raphanusol A, gentiobiose is esterified with two molecules of this substituted cinnamic acid, which are linked to the hydroxy groups at C-1 and C-4 of the gentiobiose.

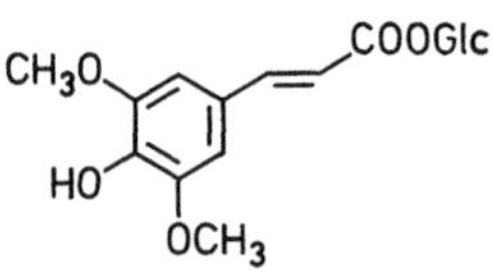

56

The lignans which are biosynthesized from cinnamic acid **54** by oxidative phenolic coupling, are also naturally occurring phenolic compounds which are widely distributed in higher plants. Their biological activities were summarized by Macrae and Towers [129].

4.2 Turgorins

It is a well known phenomenon that certain higher plants can react by autonomous movement triggered by different physical stimuli like temperature, contact, shock, light etc. and by chemicals. These nastic movements are affected by turgor changes in the cells concerned. The messengers that effect seismonastic, nyctinastic, and chemonastic movements in plants are called turgorins (cf. reviews by Schildknecht [130, 131, 132], Schildknecht and Schumacher [133] and Kallas et al. [345]). The turgorins structurally known till now have been found to be active in promoting leaf movement at hormonal concentrations (10^{-5} to 10^{-7} M) and therefore they can be regarded as phytohormone-like plant constituents. 4-*O*-(6-*O*-Sulpho-*β*-D-glucopyranosyl)gallic acid **57** represents the first structurally elucidated leaf movement factor which was isolated from the nyctinastically reacting *Acacia karroo* [134]. Later this turgorin was also detected in *Mimosa pudica*, *Acacia dealbata*, *Albizia julibrissin*, *Gleditsia triacanthos*, *Robinia pseudoacacia*, *Oxalis stricta*, *O. deppei*, *O. acetosella* and *Abutilon grandiflorum* [130, 131, 132, 135]. Several structurally closely related compounds possessing chemonastic activity could also be isolated from some species of the Fabaceae, Oxalidaceae, Malvaceae, Vitaceae, and Portulacaceae [130, 131, 132, 133, 136]. The turgorins II–VI **58–62** were identified as 4-*O*-glycosides of gallic acid, protocatechuic acid, and *p*-hydroxybenzoic acid, respectively; in each case the 6-hydroxy group of the glucosyl moiety is esterified with sulphuric acid [135, 139, 141, 142]. Turgorin VII **63** contains glucuronic acid instead of glucose. Their chemical structures are confirmed by synthesis. All these turgorins have been found by use of a special bioassay with the sensitive plant *Mimosa pudica* by estimation of leaf movement reactions after exogenous application of the corresponding turgorin.

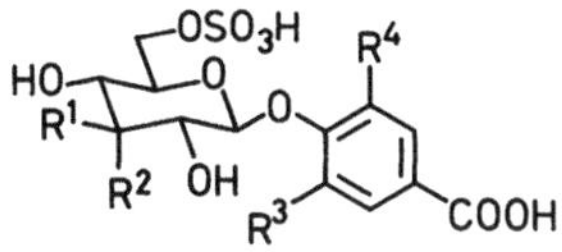

	R^1	R^2	R^3	R^4
57	OH	H	OH	OH
58	OSO_3H	H	OH	OH
59	OH	H	H	OH
60	OH	H	H	H
61	OH	H	H	OCH_3
62	H	OH	OH	OH

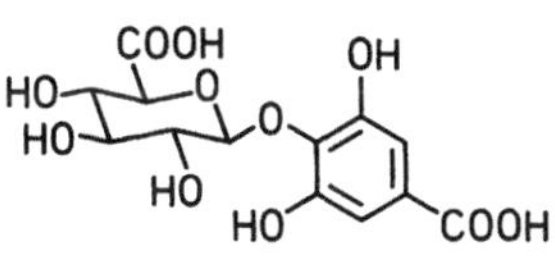

63

Recently, the effects of turgorin I 57 on stomatal movement in *Commelina communis* have also been demonstrated, but no effect was found upon leaf transpiration [137]. 2-*E*-*O*-(4-Hydroxy-cinnamoyl)-D,L-galactaric acid occurring in *Glycine max* has been proposed to be a photosensitive factor, however, recent experiments showed no effect in the bioassay with *Mimosa pudica* [138].

4.3 Stilbenes and Phenanthrenes

It has previously reviewed that some stilbenes and phenanthrenes possess plant growth regulatory properties (see [9, 10]). In subsequent work 3,3′,4,5′-tetrahydroxystilbene **64** has been detected in the heartwood of *Cassia garrettiana* as a native plant growth inhibitor [143]. Its inhibitory effects were tested on *Cucumis sativus*, *Medicago sativa*, *Raphanus sativus*, *Brassica rapa*, and *Phaseolus hirtus*.

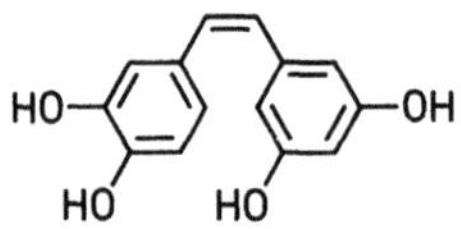

64

A highly efficient synthesis of lunularic acid **65**, previously detected in the liverwort *Lunularia cruciata* and in *Allium chinense*, has been published [144]. Testing a number of stilbene and dihydrostilbene derivatives against rice seedlings, it was found that plant growth inhibitory activity of the stilbene derivatives examined results from the presence of phenolic hydroxy groups at C-3 and C-4′.

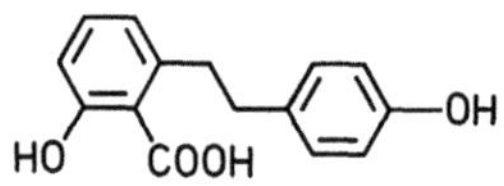

65

The structurally related phenanthrenes, aristolochic acid **66** and aristolic acid **67** from *Cocculus triolobus*, strongly inhibit seed germination of some plants [145].

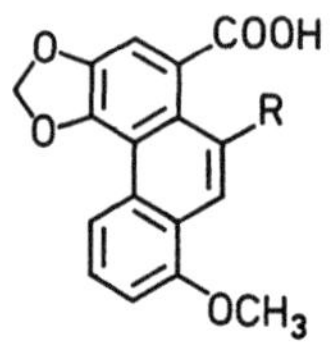

66 R = NO_2
67 R = H

4.4 Naphthoquinones

The well-known allelopathic effect of juglone **68** which occurs in the tree *Juglans nigra* in a non-toxic form as the *O*(4)-*β*-glycoside of the corresponding hydroquinone, has been confirmed by germination tests with several herbaceous and

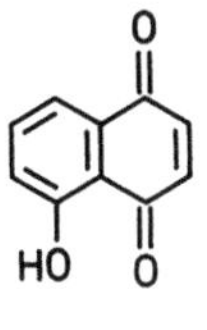

68

woody species [146]. From an unidentified *Fusarium* species several 1,4-naphthoquinones were isolated and structurally elucidated [147]. Testing their biological activity towards pollen germination of *Pinus thunbergii* the 5,8-dihydroxy-1,4-naphthoquinones **69**, **70**, **71** and **72** showed the strongest activity.

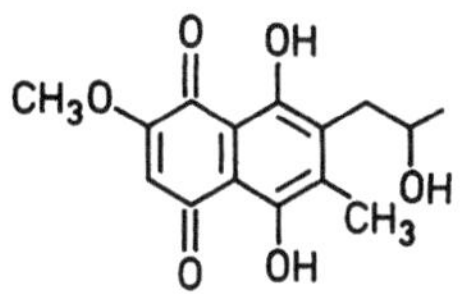

69

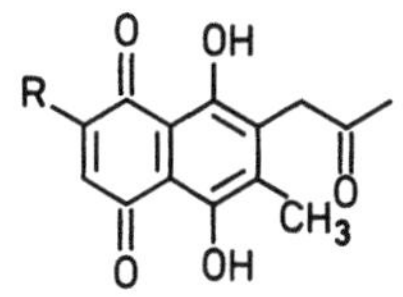

70 R = H
71 R = OCH_3

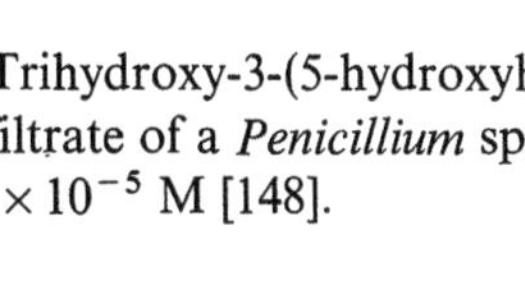

72

(*S*)-2,5,7-Trihydroxy-3-(5-hydroxyhexyl)-1,4-naphthoquinone **73**, isolated from the culture filtrate of a *Penicillium* sp. was shown to inhibit the growth of tomato callus at $1{,}5 \times 10^{-5}$ M [148].

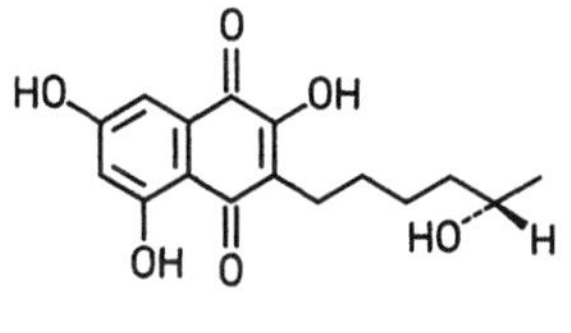

73

4.5 Coumarins

Coumarin **74** and some naturally occurring derivatives are well-known inhibitors of germination and growth (cf. [8, 9, 10, 11, 12]). Moreover, it has been reported that coumarin induced root formation and stimulated fresh weight production of hypocotyl explants of *Glycine max* [149], whereas umbelliferon **75** and scopoletin **76** from *Heracleum lanatum* highly inhibited the growth of Chinese cabbage root and bergapten **77** and sphondin **78** inhibited the growth of hypocotyl cuttings of cucumber seedlings [150].

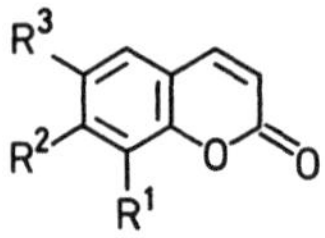

	R^1	R^2	R^3
74	H	H	H
75	H	OH	H
76	H	OH	OCH_3

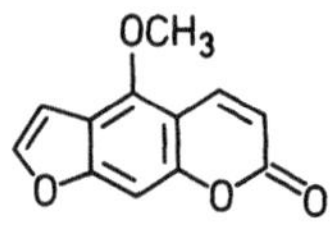

77

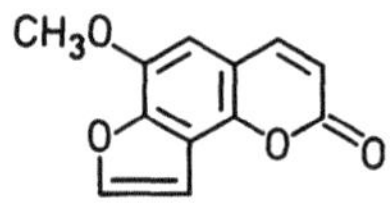

78

4.6 Flavonoids

The group of flavonoids also belongs to the phenolic compounds. Some of them exhibit plant growth properties. (+)-Afzelechin **79** and two new biflavanols **80** and **81**, isolated from *Prunus persica* roots, inhibited root growth of rice seedlings

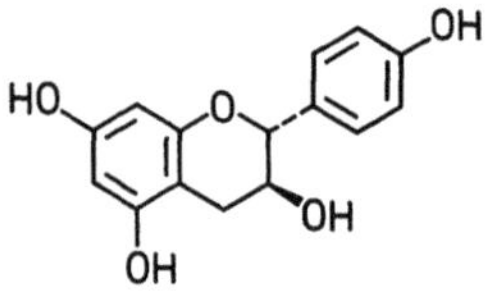

79

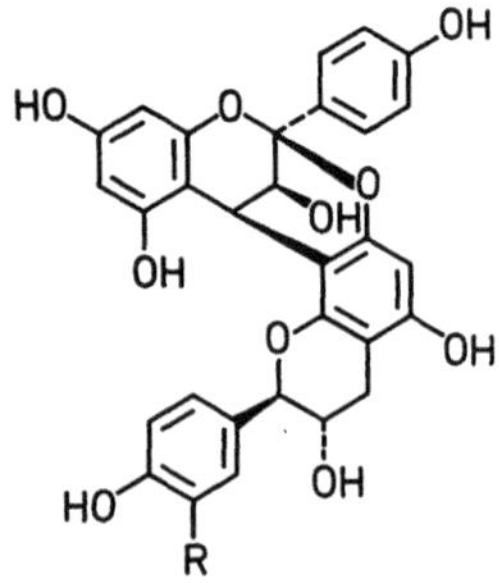

80 R = H
81 R = OH

at concentrations of 500 to 1000 ppm [151]. 5,7-Dihydroxychromone **82** from seeds of *Polygonum lapathifolium* was found to be a germination inhibitor of *Abutilon theophrasti* seeds [152].

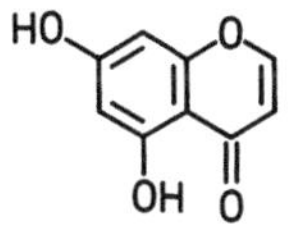

82

5 Heterocyclic Compounds

A number of naturally occurring nitrogen-, oxygen- and sulphur-containing heterocyclic compounds have been shown to possess growth regulatory properties and to affect plant growth processes. Most of them have a restricted distribution both in microorganisms and in higher plants. Some S-containing inhibitors with growth regulatory properties such as asparagusic acid from *Asparagus officinalis* have already been described in previous reviews [8, 9, 10, 11, 12].

5.1 N-Heterocyclics

A new pyrrole compound with unusual substitution has been isolated from *Pisum sativum* and structurally identified to be (*R*)-1-[3-(4,5-dihydro-2-furanone)]-5-(hydroxymethyl)-pyrrole-2-carboxyaldehyde **83** [153, 154]. This substance has been

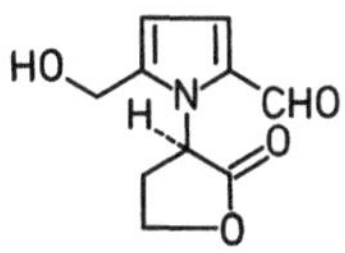

83

shown to function as a specific endogenous regulator of trigonelline induced cell arrest at the G 2 stage in pea root meristems and to interact with it at concentrations of 5×10^{-6} M. The (*S*) isomer showed no activity. The biologically active (*R*) form has an antagonistic effect of trigonelline's promotion of cell arrest (see below) and becomes the first chemically characterized substance which overrides hormonal induced cell arrest in complex tissues.

Two indole derivatives with plant growth regulatory activity have been isolated from the sponge *Dysidea etheria*, the indole-3-acetamide **84** and the hitherto unknown 4-hydroxy-5-(indol-3-yl)pentan-2,5-dione **85** [155]. It is the first report of the identification of plant growth regulators from a sponge. The indole derivative **85**, produced in *Ulosa ruetzleri*, is active in promoting root growth in lettuce seedlings. Root lengths were 120% of those controls at concentrations of 10^{-8}/ 10^{-9} M. Indole-3-acetamide is already known as a plant constituent with auxin-like plant growth regulating properties.

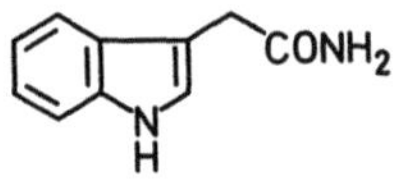

84

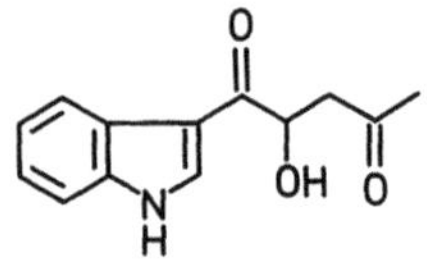

85

Harmane alkaloids have already been known to occur in various plants, mushrooms, and fungi. In the course of screening for new bioactive metabolites produced by microorganisms, norharman **86** and harman **87**, were recently isolated from the culture filtrate of an unidentified *Nocardia* sp. strain P-9 [156]. These metabolites both reduced root elongation of rice and lettuce seedlings.

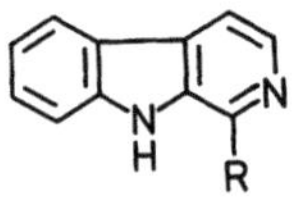

86 R = H
87 R = CH_3

Caulerpin **89** is a unique pigment produced by some species of green alga in the genus *Caulerpa* such as *C. sertularioides* and *C. racemosa* [157]. This metabolite has been structurally characterized to be a bis-indole derivative which exhibits root growth promotion activity comparable to that of indole-3-pyruvic and indole-3-acrylic acids. Caulerpin and the unesterified dicarboxylic acid **88** are novel plant growth regulators isolated from the marine biosphere.

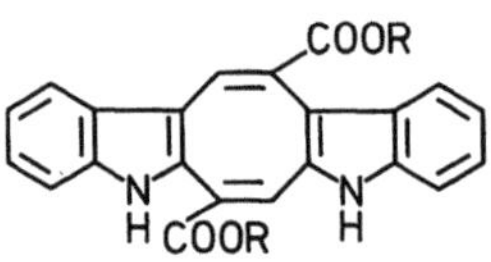

88 R = H
89 R = CH_3

Tenuazonic acid **90**, produced by a fungal pathogen of rice, *Pyricularia oryzae*, and by *Alternaria longipes* was recently shown to inhibit plant growth by inter-

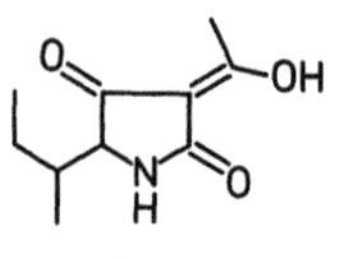

90

ferring with protein synthesis at the ribosomal level and to induce rice leaf defence reactions of *P. oryzae*-resistant varieties [158].

An endogenous germination inhibitor, recently isolated from *Triticum aestivum*, has been identified to be the common proteinous amino acid L-tryptophan **91** [159]. Tryptophan has been found to occur in leachate of oat hulls (*Avena sativa*) and to inhibit root elongation of *Avena* seedlings [297, 298]. Its *N,N*-dimethyl derivative from seeds of *Abrus precatorius* was already known as an inhibitor [160].

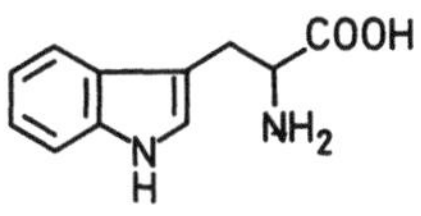

91

There are some plant growth regulators which have been found to occur in animal tissues. 1-Methylhydantoin **92** and its 5-hydroxy derivative **93** were isolated from rabbit skin tissue which has been inflamed by inoculation with vaccinia virus. These two diketopiperazines are potent rice-germination promotors [161].

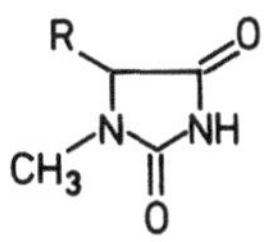

92 R = H
93 R = OH

From rabbit skin tissue extracts the hydroxypoline-derived diketopiperazines **94** and **95** have been isolated and found to be rice germination promotors at the concentrations of 10^{-6} M under dark conditions [162].

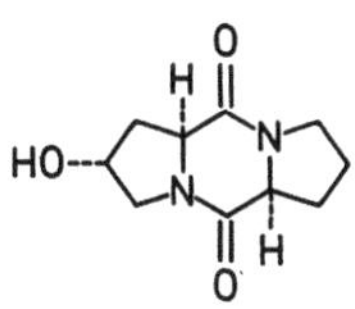

94

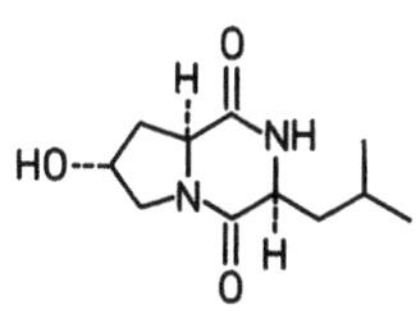

95

The extraction of the culture filtrate of the fungus *Botrytis cinerea* led to the isolation of a novel metabolite, trivially named cinereain **96** which significantly inhibited the growth of etiolated wheat coleoptiles at 10^{-3} and 10^{-4} M [163].

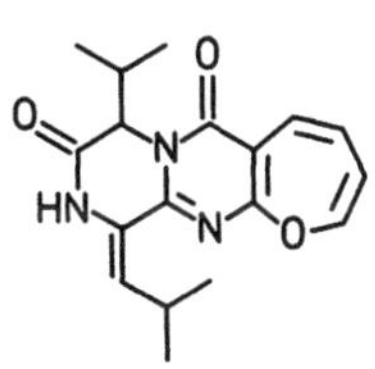

96

Furthermore, the occurrence and structural elucidation of novel 5-membered *N*-heterocyclic compounds with oxygen or sulphur as a second heteroatom, respectively, have been reported. 4-Hydroxyisoxazole **97** is a seed germination inhibitor extracted from the African folk medicinal plant *Triumfetta rhomboidea* [164]. This inhibitor is named triumferol. Its lettuce seed antigermination activity was 100% at 100 ppm.

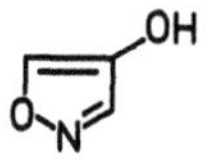

97

Light-exposed seedlings of radish, *Raphanus sativus*, contain a plant growth inhibitor which has been characterized as 2-thioxothiazolidine-4-carboxylic acid **98**. Its growth inhibitory properties have been shown by bioassays with hypo-

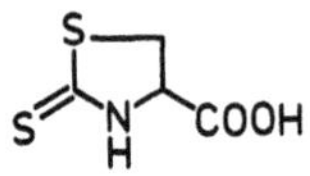

98

cotyls of etiolated radish and lettuce seedlings [165]. Raphanusanin, a neutral growth inhibitor also from light-exposed radish seedlings, has been identified as 3-methoxy-4-methylthio-2-piperithione **99** [166, 367]. Raphanusanin inhibited the hypocotyl growth of etiolated radish and lettuce seedlings at concentrations higher than $1{,}5 \times 10^{-6}$ M. It may play a role in light inhibition of hypocotyl growth of radish seedlings as well as raphanusol A and B.

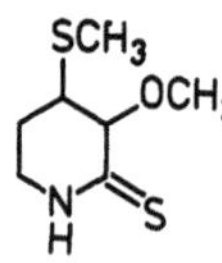

99

The different biological activities of the benzoxazinones, long known to be constituents of several Gramineae species, have been very recently reviewed [167]. The plant growth inhibitory effects of 2,4-dihydroxy-1,4(2*H*)-benzoxazin-3-one **100** and its degradation product 2(3*H*)-benzoxazolinone **101** both occurring in

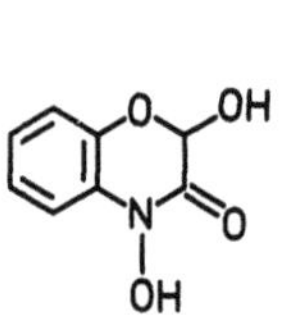

100

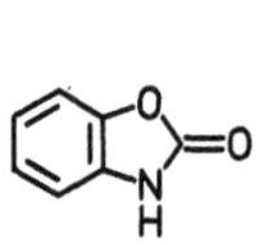

101

rye, *Secale cereale*, and in *Acanthus mollis* seeds have been reported [168, 169, 170].

The betaine trigonelline **102** was first isolated from *Trigonella foenum-graecum* and was subsequently detected in many plants and other biological sources. It usually represents the most abundant compound of the pyridine nucleotide cycle for the production of NAD.

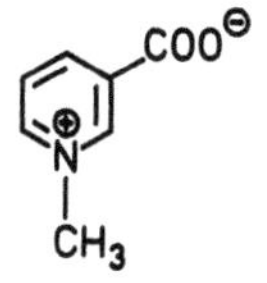

102

Trigonelline is contained in cotyledons of *Pisum sativum* and transported to roots and shoots after germination. It blocks mitosis at the G 2 stage (Gap-2) both in cultured roots and shoots during normal cell differentiation of *Pisum sativum* and various legumes [171, 172, 173, 174, 175, 299, 300]. Trigonelline has been proposed as a hormone that controls the cell cycle in plants, at least of legumes. Nicotinic acid **103** and nicotinamide **104** known to be precursors of trigonelline **102**, are active in promotion of cell arrest only because they are converted to trigonelline [176, 299]. Experiments with several synthetic analogues of trigonelline have shown that only *N*-methylquinolinic acid and *N*-methylnicotinamide induced cell arrest in G-2 [177]. These data further support a specific role for trigonelline in induction of cell arrest in roots of legumes.

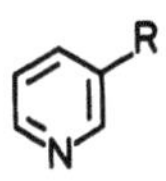

103 R = COOH
104 R = $CONH_2$

In addition to benzoic acid **31** and salicylic acid **32** several other flower-inducing factors were shown to be present in *Lemna* species as *L. gibba* and were identified as nicotinic acid **103**, nicotinamide **104** and L-pipecolinic acid **104a** [292, 293, 294, 295]. These compounds induced flowering of *Lemna paucicostata*, but they were more active in the presence of benzyladenine.

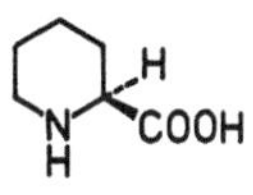

104a

The germination of lettuce was inhibited by mixtures of quinolizidine alkaloids from *Lupinus mutabilis* seeds and aerial parts of *L. polyphyllus* [178]. It is assumed that lupin alkaloids such as lupanine **105** may be potential components of plant-plant interaction. The aspects of allelopathy in *Lupinus albus* was discussed.

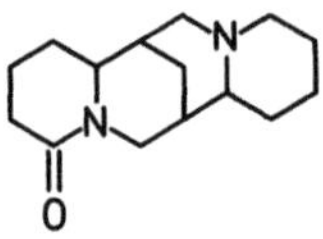

105

Some phenylethylamines and tetrahydroisoquinolines that have been found to occur in peyote and other cacti showed strong inhibition in a bean internode bioassay and a *Sorghum* seedling bioassay [179].

5.2 O-Heterocyclics

3-Carboxy-2,4-diphenyl-but-2-enoic anhydride **106** and asperlin **107** produced by the fungus *Aspergillus nidulans* were shown to accelerate elongation of radish

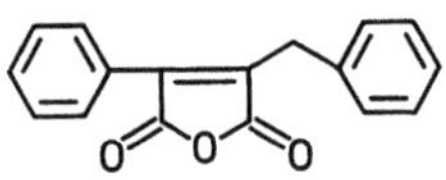

106

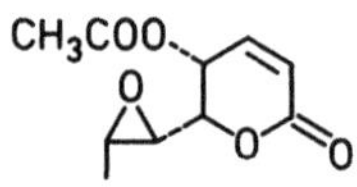

107

and lettuce seedling roots [180]. 6-Pentylpyran-2-one **108** was first isolated from *Trichoderma viride* and *T. harzianum*. It significantly inhibits the growth of etiolated wheat coleoptiles at concentrations of 10^{-3} M [181]. Structurally related compounds, 6-(1, 2-dihydroxy-pentyl)-4-methoxypyran-2-one **109** and its 5,6-dihydro derivative **110** are produced by *Penicillium citreo-viride*. Supplied together with GA_3, compound **109** enhanced the stimulative effect of GA_3, but alone

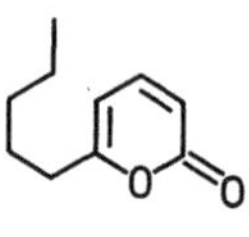

108

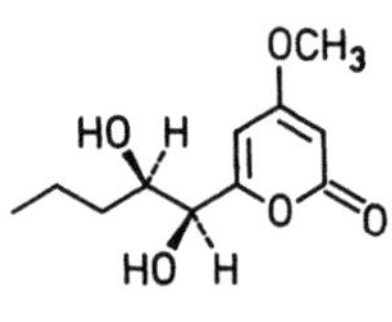

109

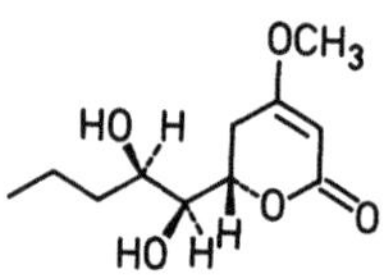

110

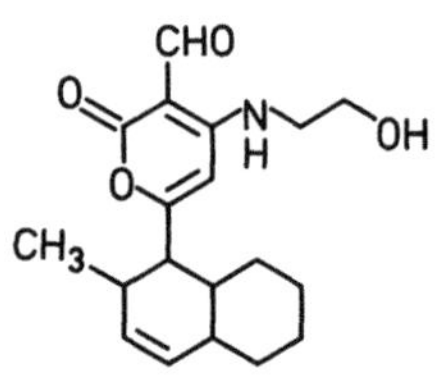

111

it was inactive. The dihydro derivative influenced the elongation in the presence or abscence of GA_3 [182]. The extraction of the culture filtrate from *Alternaria solani* led to the isolation of an α-pyrone derivative **111** which has shown plant growth inhibition in bioassays with lettuce and potato [183].

Parasorbic acid **112** has been isolated from ripe fruits and seeds of *Sorbus aucuparia* [46]. This lactonic compound inhibited germination of seeds from *Amaranthus caudatus* and *Lepidium sativum* at concentrations of 5×10^{-4} M.

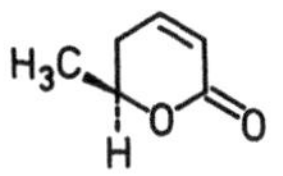

112

Dormant seeds of *Rosa canina* contain new α-pyrones **113** and **114** which act as potent germination inhibitors of *Amaranthus caudatus* seeds [184]. One inhibitor was identified as dimethyl 3-methyl-2*H*-pyran-2-one-4,6-dicarboxylate **113**, the other is a structurally related compound.

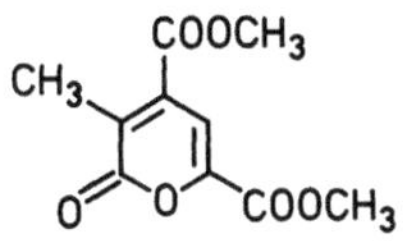

113

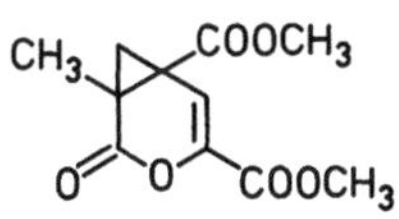

114

Three derivatives of 2,3-dioxabicyclo[4,4,0]decane **115**, **116**, **117** are constituents of *Eucalyptus grandis* leaves and of some other myrtaceous genera (cf. reviews [8, 10]). By using $^{18}O_2$ it was demonstrated that these *Eucalyptus* constituents are really natural products and not artifacts. These bicyclic compounds, designated

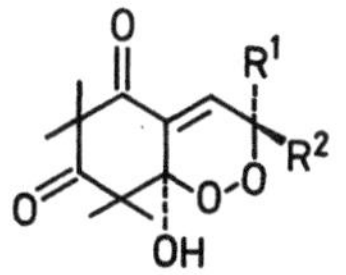

	R^1	R^2
115	CH_3	C_2H_5
116	C_2H_5	CH_3
117	CH_3	CH_3

G-inhibitors, have been shown to control, over a wide range of physiological functions in plants, for example stimulation of rooting of mung bean cuttings at 10^{-5} to 10^{-6} M and inhibition of elongation in an *Avena* bioassay [185]. Recently these native plant growth regulators were prepared by a biomimetic synthesis [186].

The phytopathogenic fungus *Neocosmospora vasinfecta* produces two bicyclic compounds which are named neovasinone **118** and neovasinin **119** [187, 188]. These α-pyrones remarkably accelerated the root elongation of lettuce seedlings.

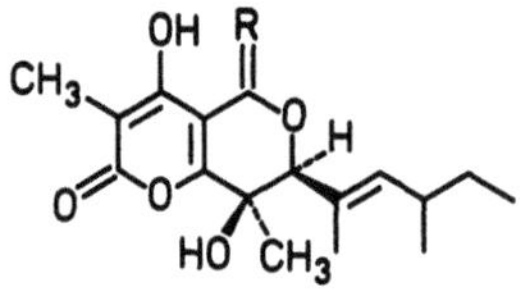

118 R = O
119 R = H_2

6 Amino Acids and Peptides as Metal Chelating Agents

Some unusual, non-proteinous amino acids derived from azetidine-2-carboxylic acid have been found to possess metal chelating properties. Because of the interesting biological activity and the essential functions in the mineral metabolism of plants this new class of plant effectors will also be discussed here.

The amino acid (—)-nicotianamine **120** has first been isolated from leaves of *Nicotiana tabacum* [189] and was shown to occur in *Fagus silvatica* [190], in several species of the Solanaceae, Gramineae and Liliaceae [191] and in some further plants [192]. Nicotianamine was chemically characterized as (2*S*:3′*S*:3″*S*)-*N*-[*N*-(3-amino-3-carboxypropyl)-3-amino-3-carboxypropyl]-azetidine-2-carboxylic acid. Later nicotianamine was identified as the "normalizing factor" which restores

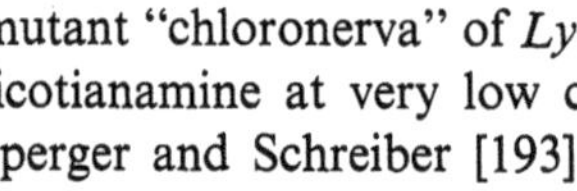

120

defects of chlorophyll synthesis, growth and development of the nicotianamine-auxotroph mutant "chloronerva" of *Lycopersicon esculentum* by exogenous application of nicotianamine at very low concentrations (1 nmol per seedling) (see reviews Ripperger and Schreiber [193]; Prochazka and Scholz [194]; Schreiber [195]). Due to a metabolic block in the biosynthetic pathway of nicotianamine the tomato mutant "chloronerva" does not contain this "normalizing factor". Thus, the mutant shows a disturbed iron metabolism leading to an excessive iron uptake by the roots and an irregular iron distribution within the leaves. Very recently the regulation of iron uptake and the possible functions of nicotianamine in higher plants have been extensively reviewed [196].

Nicotianamine seems to be of general distribution in the plant kingdom (cf. [193, 194 195]). It forms stable complexes in a 1:1 range with iron (II) and other bivalent transition-metal ions, such as Cu, Ni, Co, Zn, and Mn [197]. The biological activity of this potent chelator correlates with the coordination behaviour as recently shown by testing synthetic nicotianamine analogues [198].

To our present knowledge nicotianamine acts as an endogenous chelating agent within the plant and plays an important role in the short-distance inter- and intracellular transport and metabolism of bivalent metal ions. According to the present definitions nicotianamine can be regarded as a cytometallophore.

Some other naturally occurring hydroxy substituted iminocarboxylic acids, such as mugineic acid **121** from *Hordeum vulgare* [199], 2′-deoxymugineic acid **122** from *Avena sativa* and *Triticum aestivum* [200, 201], 3-hydroxymugineic acid from

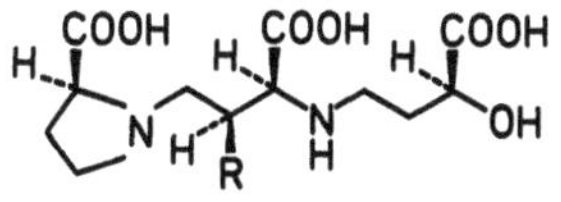

122 R = H
121 R = OH

Secale cereale [202] and avenic acid A **123** and B **124** from *Avena sativa* [203, 204], are closely structurally related to nicotianamine. The mugineic acids have also been shown to be potent chelators responsible for the iron (III) ions uptake by the roots [205, 206, 207]. It may be that these amino acid are metabolites of nicotianamine.

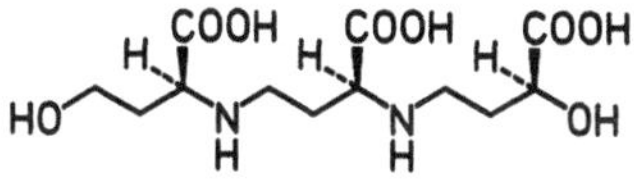

123

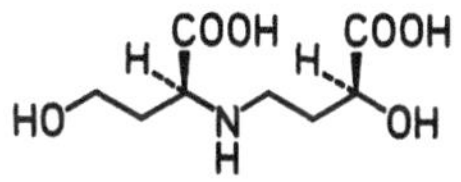

124

The synthesis of nicotianamine and the other structurally related amino acids as well as of various synthetic analogues has also been reported (cf. reviews [193, 195] and references therein).

The relationship between siderophores produced by microorganisms and plant growth and disease has been the subject of recent reviews [208, 209]. The siderophores, therefore, are excluded here.

In this connection it should be mentioned that very recently the occurrence of heavy-metal binding phytochelatins in higher plants and plant cell suspension cultures upon exposure to certain heavy metals was described [210, 211, 212, 213]. Phytochelatins are inducible peptides which consist of repetitive γ-glutamylcysteine units with a carboxyl-terminal glycine and range from 5–17 amino acids in length. These peptides are capable of binding heavy metal ions via thiolate coordination. Such phytochelating peptides were also found in the *Chlorella fusca* and other classes of algae after treatment with cadmium and other toxic metal ions [214]. Phytochelatin formation is specifically induced in plants by a variety of toxic metal ions, such as Cd^{2+}, Zn^{2+}, Pb^{2+}, Ag^{+}, Cu^{2+}, and Hg^{2+}. All data so far obtained indicate that phytochelatins are responsible for the cytoplasmic detoxification of heavy metals in algae and higher plants (recent review see [368]).

7 Polyamines

Aliphatic diamines and polyamines are of wide occurrence in both animals and plants. Numerous investigations have shown the involvement of spermidine **126** and spermine **127** and their biogenetic precursor putrescine **125** in the regulation of various plant growth and developmental processes.

125

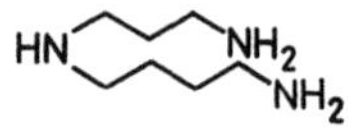

126

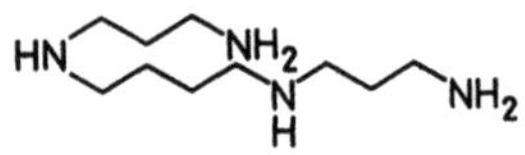

127

Polyamines are biogenetically derived from L-arginine which is transformed to putrescine via either the ornithine or agmatine pathway. Addition of a methionine-derived aminopropyl group either to putrescine or spermidine leads to the triamine, spermidine, and the tetraamine, spermine, respectively. The enzymes catalyzing the conversion of arginine to the polyamines have been studied in detail. In plants, polyamines may be free, bound to membranes, or conjugated to small molecules such as hydroxycinnamic acids.

There are a high number of experimental studies dealing with the different effects and physiological activities of polyamines in higher plants and other biological systems. Most of the relevant literature has been well covered in previous reviews. The role of polyamines in the control of cell proliferation and differentiation was described by Heby [215]. The biochemistry of polyamines in eukariotes has been reviewed by Pegg [216]. For a discussion of the physiology and biochemistry of polyamines and their involvement in plant growth and in adaption to stress, the reader is referred to surveys of Flores et al. [334], Evans and Malmberg [335], McCann et al. [217], Slocum et al. [218], Tabor and Tabor [219, 220] and Smith [221] and references therein. A special issue of "Plant Growth Regulation" (Vol. **3** (3/4) 1985) is devoted to 17 papers on polyamines in plants. A review entitled "Effects of Polyamines and Polyamine Inhibitors on Growth, Differentiation and Organogenesis in Plant Cell Cultures" was published by Serafini-Fracassini [222].

In addition to these comprehensive articles which give extensive information in the field of polyamine research, some recent papers are discussed very briefly.

The influence of some synthetic inhibitors on the enzymes of plant polyamine biosynthetic pathway was studied in leaf explants of *Nicotiana xanthi* [223] and in calli of *Abutilon theophrasti* and tomato seedlings [224, 225]. Similar experiments were carried out with nor-spermidine which inhibited growth of potato plantlets [226].

Exogenous application of spermidine caused floral bud initiation in tissue cultures of *Nicotiana tabacum* indicating the florigenic action of spermidine, at least in the tobacco system used [227]. The polyamine biosynthesis during vegetative and floral bud differentiation has also been investigated in thin layer tobacco tissue cultures [228]. Moreover, it has been shown that conjugated polyamines, expecially putrescine, seem to be involved in both reproductive and vegetative phases of cultured tobacco thin cell layers [229].

8 Iridoids

A few iridoids have been found to inhibit seed germination. For example, asperulosidic acid and its deacetyl derivative were isolated from *Galium aparine* and shown to suppress seed germination of *G. aparine*, alfalfa, lettuce and crabgrass and the growth of radicle and hypocotyl [230].

Geniposide **128** and geniposidic acid **129** together with their aglycones **130** and **131** isolated from *Garrya elliptica* inhibited the elongation of wheat coleoptiles [231]. Fruits of *Gardenia jasminoides* contain geniposide **128** and genipin gentiobioside **132** as germination inhibitors [232].

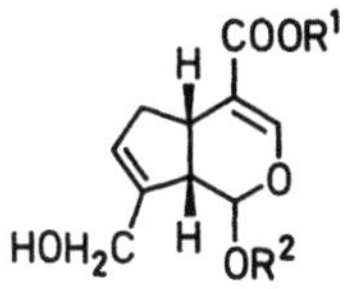

	R^1	R^2
128	CH_3	Glc
129	H	Glc
130	CH_3	H
131	H	H
132	CH_3	β-gentiobiose

Plumieridine **133** obtained by enzymatic cleavage of the iridoid glucoside plumieride from *Plumeria acutifolia* showed plant growth inhibitory effects [233].

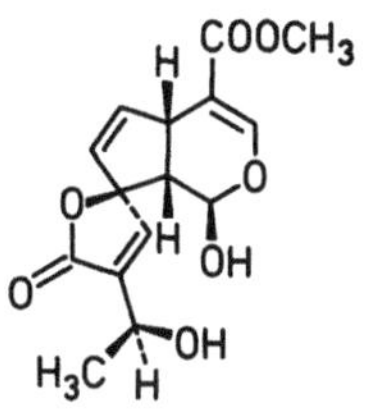

133

9 Terpenoid Compounds

A high number of terpenoid compounds are known which affect plant growth and development. Many of them have been the subject of earlier reviews (cf. [8, 9, 10, 12]). Thus, only a few examples are illustrated here.

Some sesquiterpenoids, closely structurally related to the well-known phytohormone abscisic acid, were found to be produced by the fungus *Cercospora cruenta*, such as 2*Z*,4*E*-γ-ionylideneethanol **134**, 2*Z*,4*E*-γ-ionylideneacetic acid **135** and its 1′,4′-dihydroxy derivative **137** [234, 235].

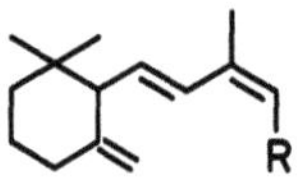

134 R = CH_2OH
135 R = COOH

Another analogue of abscisic acid is 2*Z*,4*E*-4′-hydroxy-γ-ionylideneacetic acid **136** which has been isolated from a *Stemphylium* sp. [236]. These compounds have been shown to be biosynthetic intermediates of abscisic acid and to exhibit inhibitory properties, but their activities were less than those of abscisic acid. It is not yet clearified whether these sesquiterpenoid compounds act *per se* or only after conversion to abscisic acid which is inhibitory within the plant used in the bioassay.

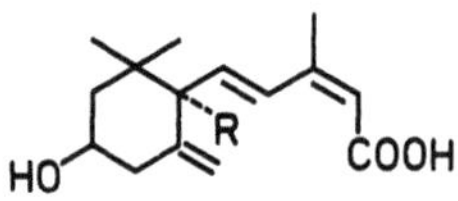

136 R = H
137 R = OH

(—)-Isobicyclogermacrenal **138** and (+)-vitrenal **139** are plant growth inhibitors isolated from the liverwort *Lepidozia vitrea* [237, 238, 239]. (+)-Vitrenal with its novel sesquiterpenoid carbon skeleton completely inhibits the growth of leaves and roots of rice seedlings at 10^{-4} M. The synthesis and biological activity of (—)-vitrenal have also been reported [240].

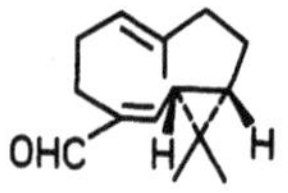

138

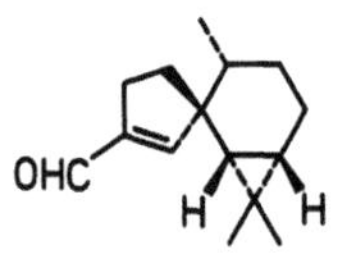

139

Pinthunamide **139a** is a newly found tricyclic sesquiterpene amide which is produced by the fungus *Ampulliferina* sp. [296]. It accelerated the root growth of lettuce seedlings.

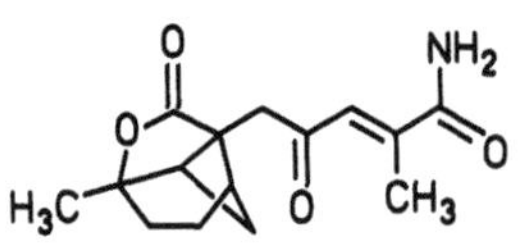

139 a

Another liverwort, *Plagiochila semidecurrens*, contains the sesquiterpene (+)-9α-acetoxyovalifoliene **140** and some ovalifoliene-related constituents which showed strong inhibition of rice seedling growth [241]. Plagiochiline A and C and ovalifolienal from *Plagiochila ovalifolia* were also shown to be growth inhibitoring [242].

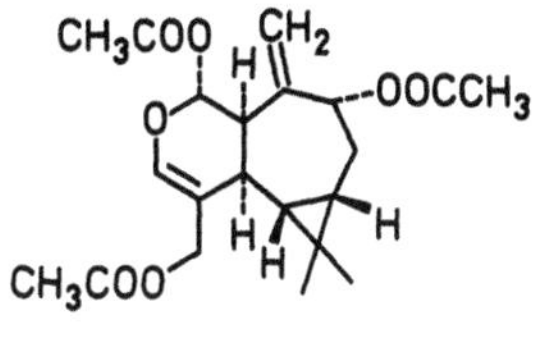

140

Peruvin **141** and ambrosic acid **142** are growth inhibitory compounds from *Ambrosia artemisiifolia* [243]. The fungus *Dendrostilbella* sp. has been reported to produce trichodermin **143** which significantly inhibited the growth of etiolated wheat coleoptiles [244]. Isoinunal **144**, a newly discovered potent plant growth regulator from *Inula racemosa*, has been shown to initiate root formation of hypocotyl cuttings from *Phaseolus aureus* [245]. Some sesquiterpene lactones, such as

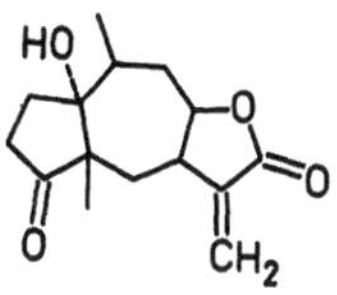

141

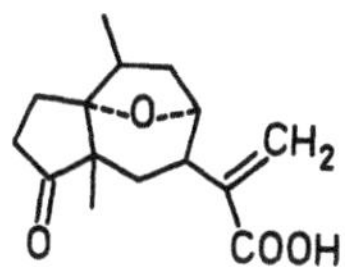

142

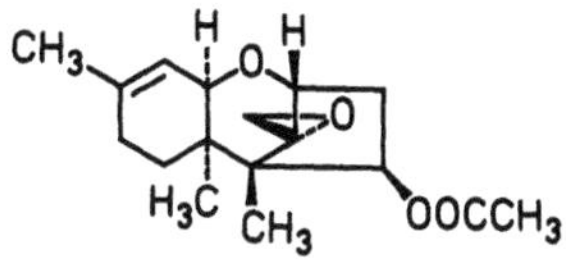

143

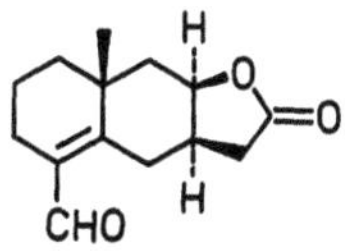

144

niveusin C **145** and 15-hydroxy-3-dehydro-deoxyfruticin **146** (ang = angelic acid) were isolated from *Helianthus annuus* and shown to inhibit auxin-induced elongation growth of oat coleoptile segments and *Helianthus annuus* hypocotyl segments [246, 247].

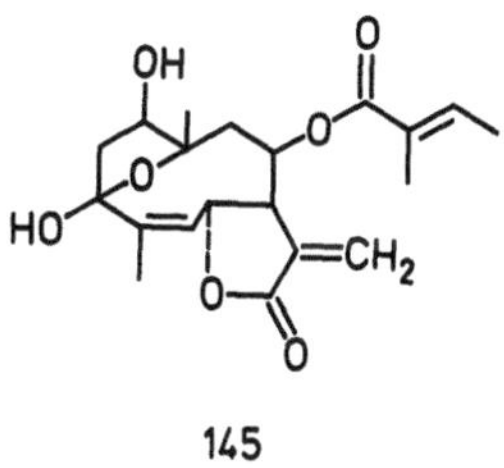

145

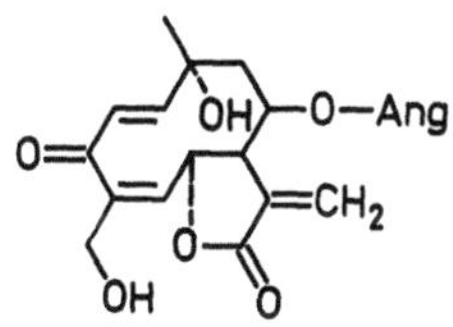

146

Strigol **147** is known as a highly potent seed germination stimulator for *Striga lutea*. The preparation of (+)-strigol [365], a strigol precursor [285] and some synthetic analogues [248] has been published. A sensitive bioassay for strigol using *Striga asiatica* seed germination was described [249].

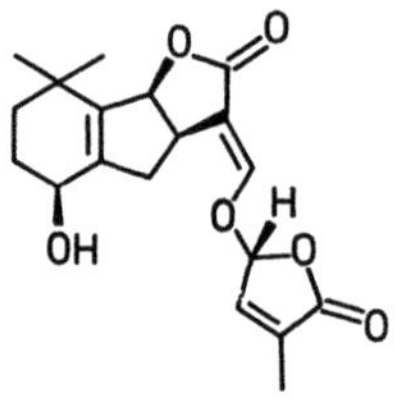

147

Khusinoloxide **148** and other naturally occurring cadinane terpenoids, such as *epi*-khusinol acetate, have been tested as plant growth regulators and found to stimulate root initiation with hypocotyl cuttings of *Phaseolus aureus* seedlings [250].

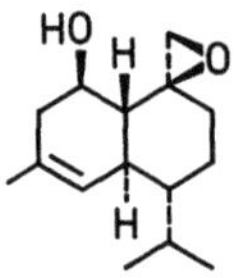

148

Substances inducing antheridia formation have been termed antheridiogens. These hormone-like compounds with gibberellin-like properties and closely related in structure to the gibberellins have a direct influence on the sexual morphology of developing gametophytes, including formation of antheridia, and spore germination in certain species of ferns. Antheridiogen activity has been detected in many different genera of fern, but now the first structures of these biologically active compounds have been elucidated.

From *Anemia phyllitidis* the gibberellin-like diterpene antheridic acid **149** was isolated and identified in its chemical structure and absolute configuration [251]. The structure was confirmed by a total synthesis of the racemate [252] and by a partial synthesis starting from gibberellin A_7 [253, 262]. Antheridic acid has also been detected in *Anemia rotundifolia* and *A. flexuosa* [254].

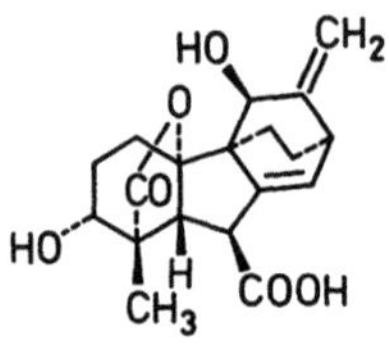

149

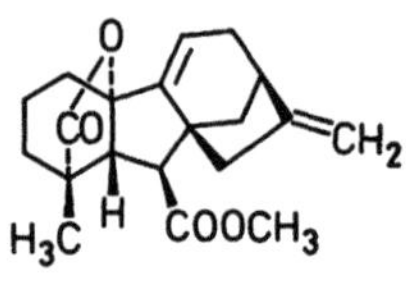

150

A structurally related gibberellin-like C_{19}-compound **150** was isolated from *Lycodium japonicum* and characterized as 9,11-didehydrogibberellin A_9 methyl ester (methyl ester of gibberellin A_{73}) [255]. In *L. japonicum* it exhibited high activity in inducing antheridial formation at 10^{-15} M and dark spore germination at 10^{-12} M. Gibberellin A_9 methyl ester occurs also in *L. japonicum* and was described as antheridiogen [256].

Very recently a novel antheridiogen **151**, isolated from gametophytes of the fern *Anemia mexicana*, was structurally elucidated and shown to possess a novel *ent*-9,15-cyclogibberellane skeleton [257, 258, 259, 260]. The synthesis of this antheridiogen has also been achieved [261].

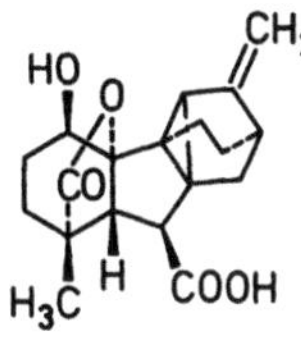

151

In earlier reviews mentioned above, nor- and bisnor-diterpene dilactones have been summarized and described as endogenous plant growth regulating constituents. Later some other structurally related compounds were also under investigation and were shown to possess significant plant growth regulating activity [263, 264, 265, 266, 267]. The synthesis and biological activities of α-methylene-γ-butyrolactones were reviewed by Hoffmann and Rabe [268]. For detailed information, the reader is referred to these papers and previous reviews.

10 Steroids

Several steroids produced by plants have been shown to exhibit plant growth regulating activities. Studying the role of 44 steroids on plant growth and development in various bioassay systems it has been demonstrated that glucocorticoids mainly are growth stimulating substances, whereas mineralocorticoids are mostly growth-inhibiting [269]. Amasterol from *Amaranthus viridis* roots causes inhibition of seed germination and growth of seedlings [270]. Vitamin D_2 and D_3 have been under further investigation and were shown to promote adventitious root formation in stem cuttings of *Phaseolus aureus* [271, 272] and cuttings of *Populus tremula* [273, 274, 275, 288].

At present, the brassinosteroids, representing the most important group of plant steroids with plant growth regulatory properties, receive much attention as native plant growth regulators. They are discussed by Marquardt and Adam in a separate chapter.

11 Oligosaccharins

Recently, it was shown that certain fragments from both fungal and plant cell walls are very highly effective elicitors which induce the *de novo* synthesis of enzymes catalyzing the synthesis of phytoalexins in higher plants (cf. reviews of Albersheim and Darvill [276]; Dixon [277]; Ebel [278]; Keen [369] and Ryan [279, 280]). Some active elicitors, obtained from fungal and plant cell wall preparations either by partial acid hydrolysis or degradation with glucanase or endopolygalacturonase, could be identified as oligosaccharides built up from glucose and galactose, respectively. For example, a highly potent elicitor-active fragment from mycelial cell walls of *Phytophthora megasperma* was characterized as hexa-(β-D-glucopyranosyl)-D-glucitol [281]. Defined fragments of cell-wall oligosaccharides that have been found to give chemical signals with high regulatory properties in plants are generally termed oligosaccharins. First results indicate that oligosaccharins can regulate not only triggering of plant defences against phytopathogens but also regulate growth and differentiation of plants (see Albersheim and Darvill [276]; Darvill et al. [282]). It has been reported that oligosaccharins may be implicated in the regulation of flowering and morphogenesis in thin cell-layer explants of tobacco [283]. Growth of barley coleoptiles was found to be regulated by minor cell wall components such as galactose and mannose [284]. Nevertheless, the exact role of oligosaccharins in plant morphogenesis and their possible interactions with the classic phytohormones remains to be established.

12 Miscellaneous Compounds

Furthermore, some fungal metabolites with plant growth inhibiting properties should be mentioned (see also [15]).

The fungus *Macrophomina phaseolina* is known to produce a non-specific exotoxin which inhibits the seed germination of *Phaseolus mungo* [301]. The structure of this phytotoxin, phaseolinone, has been established as **152** [302]. Cladosporin **153**, isolated from the fungus *Aspergillus repens*, has been found to inhibit signi-

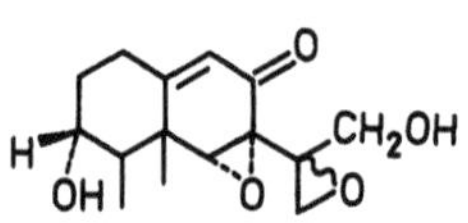

152

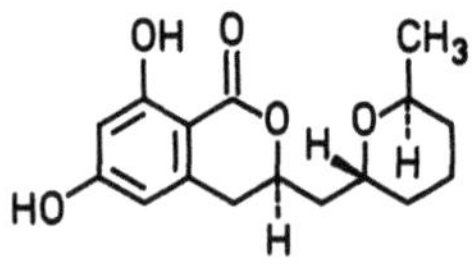

153

ficantly the growth of etiolated wheat coleoptiles at 10^{-3}, 10^{-4}, and 10^{-5} M [303]. From the culture filtrate of *Phomopsis sojae* the cytochalasin mycotoxins **154–157** were isolated and found to show plant growth inhibition in a wheat coleoptile bioassay [304, 305].

Aspergillus terreus produced the diketopiperazine **158** (acetylaranotin), the bisdethiodi(methylthio)acetylaranotin **159** and terrein **160** [306]. All these fungal metabolites reduced root elongation of lettuce seedlings.

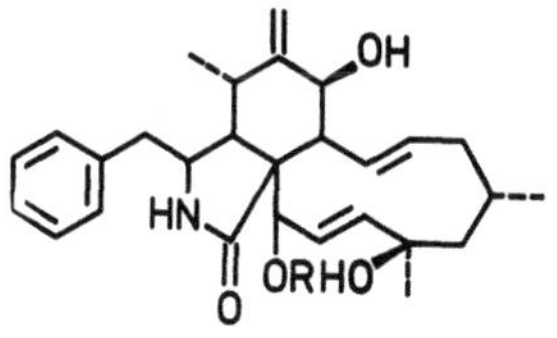

154 R = $COCH_3$
155 R = H

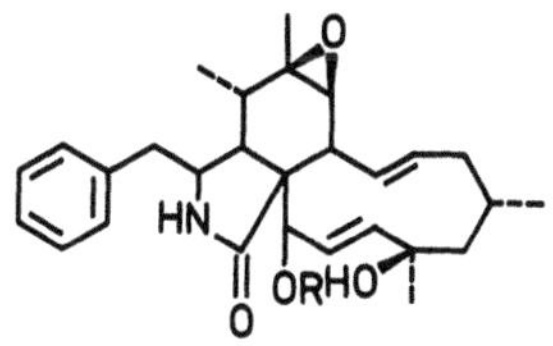

156 R = $COCH_3$
157 R = H

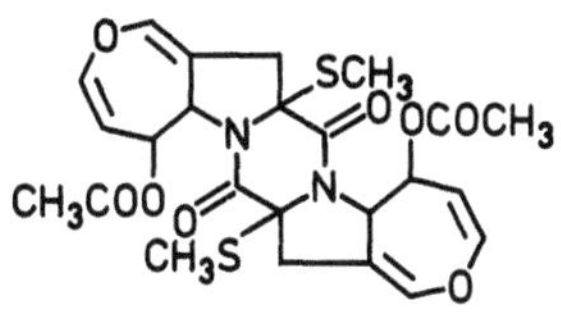

158

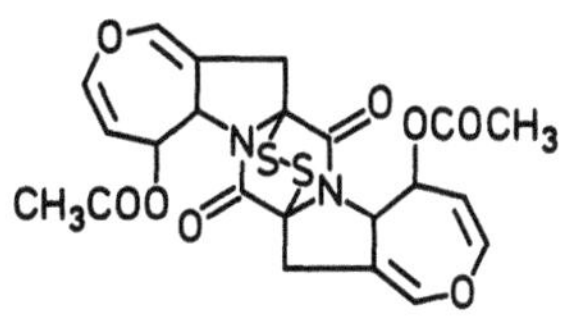

159

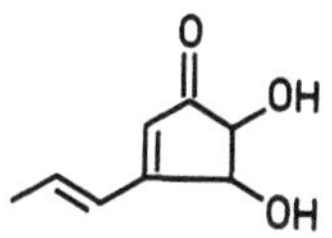

160

13 Addendum (Added in Proof)

While this review was being printed reports of a number of novel plant growth regulating substances have been published.

A review of plant growth regulators and insect control agents from marine organisms was given by Cardellina et al. [308].

Some volatile aliphatic 2-ketones and alcohols in the range C_4 to C_{11} have been discovered in *Amaranthus palmeri* and they inhibited significantly the germination of onion, carrot, tomato, *A. palmeri* and some other plants [309, 310]. Such methyl ketones and alcohols are widespread in ants as alarm-defense compounds. Due to this chemical similarity, certain of them are suggested as acting additionally as seed germination inhibitors in the nests of granivorous species. Recently, it has been found that 4-methyl-3-heptanone and some structurally related compounds, isolated from myrmicine ants, caused strong inhibition of the germination of the grass *Eragrostic curvula* [311]. These results are in agreement with the assumption that some ant alarm-defense secretion may also control seed germination inhibition.

Very recently, the *O*-*β*-D-glucopyranoside of tuberonic acid **161** has been isolated from potato leaves as a tuber-inducing substance [312, 339]. Jasmonic acid **21**, its methyl ester **22** and cucurbic acid **23** also possess tuber-inducing activity [312].

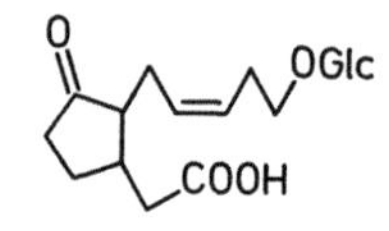

161

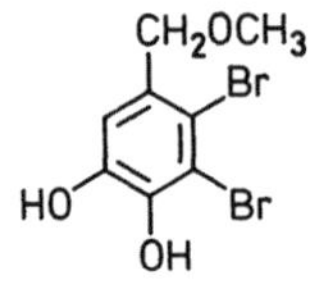

162

Methyllanosol **162** from the marine algae *Odonthalia washingtoniensis* and *O. floccosa* stimulated hypocotyl growth and root growth of lettuce and rice seedlings [313]. From the aquatic plant *Acorus gramineus* some phenylpropanes like **163–165** have been isolated and found to possess antialgal activity against several species of green and blue-green algae [314].

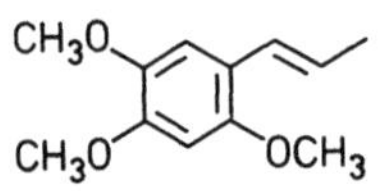

163

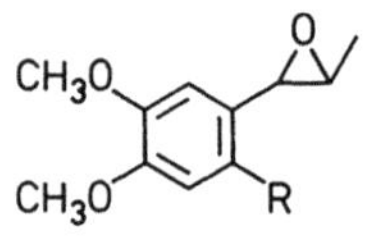

164 R = H
165 R = OCH_3

From *Artemisia capillaris* capillarol **166** was isolated as root growth promoting substance [315, 316]. It was identified as methyl 3-(3-methylbut-2-enoyl)-4-hydroxycinnamate. A plant growth inhibitor from immature seeds of corn was

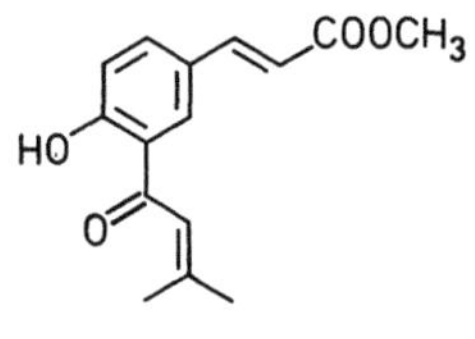

166

characterized as (*S*)-(+)-1-(*p*-hydroxy-*trans*-cinnamoyl)glycerol **167** [317]. It inhibited the inclination of the rice lamina joint and root elongation of Chinese cabbage seedlings. A lateral root inducing compounds, isolated from the bacterium *Erwinia quercina*, was shown to be 4,5-dihydroxy-6-phenylhexanoic acid **168** [318].

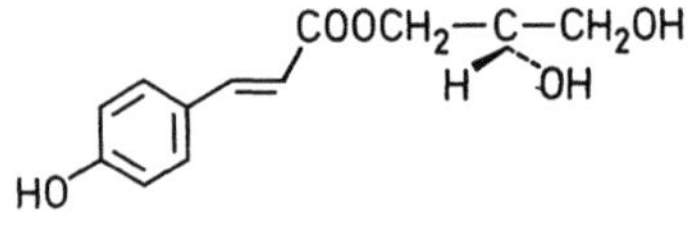

167

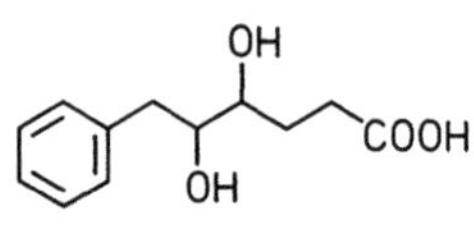

168

The dihydrostilbene phytoalexins batatasin IV **169** and dihydropinosylvin **170** from infected or chemically stressed tubers of *Dioscorea alata* were found to inhibit the germination of seeds and root elongation of young seedlings of *Sorghum bicolor* [319].

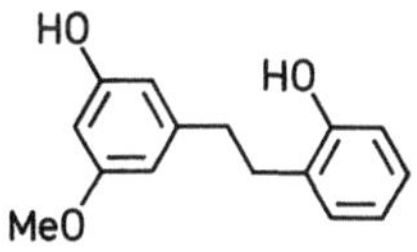

169

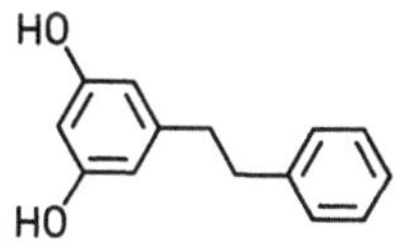

170

3,7-Dimethyl-8-hydroxy-6-methoxyisochroman **171** was isolated as a metabolite of *Penicillium corylophilum* [320]. It inhibited significantly the growth of etiolated wheat coleoptiles. Spiciferone A **172** is a novel plant growth inhibitor produced by the plant pathogenic fungus *Cochliobolus spicifer* [321]. It contains an unusual bicyclic unit composed of a fully substituted γ-pyrone and a cyclohexenone.

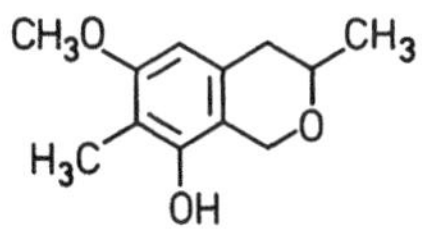

171

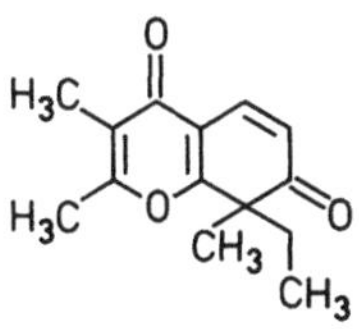

172

Acremoauxin A **173**, produced by *Acremonium roseum*, was identified as the D-arabitol ester of 2-(3-indolyl)propionic acid [333]. This novel auxin derivative has been shown to be a potent plant growth inhibitor. Another strong plant growth inhibitor from the African tree *Baillonella toxisperma* was identified as 3-hydroxyuridine **174** [322]. It inhibited both hypocotyl and root growth of cucumber and radish seedlings and root growth of rice seedlings. It is postulated that it may be involved in the allelopathy of this tree in the tropical rain forest of Cameroon.

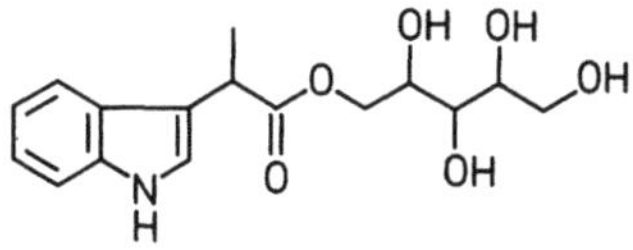

173

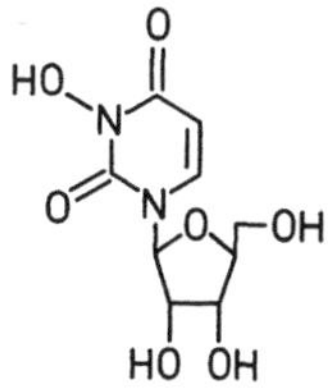

174

The correlation between metal complex formation and biological activity of nicotianamine **120** has been studied in more detail [323]. The results obtained are in agreement with the assumption of a transition-metal ion carrier function of nicotianamine in plants. Studying the influence of nicotianamine and iron supply on formation and elongation of adventitious roots in hypocotyl cuttings of tomato plants (mutant 'chloronerva') it was confirmed that nicotianamine is involved in the intracellular transport of iron [324]. Moreover, the role of nicotianamine in the regulation of iron metabolism and iron deficiency response mechanism has been discussed [325]. Studies with D- and L-forms of avenic acid A **123** and deoxydistichonic acid indicated the existence of stereospecific Fe-uptake mediated by phytosiderophores in gramineous plants [326].

In connection with the occurrence and function of phytochelatins it has been reported that more than 200 individual species of lower and higher plants form phytochelating heavy-metal complexes after exposure to cadmium ions [327].

The sesquiterpene lactone isoalantodiene **175** from roots of *Inula racemosa* was shown to be a particularly potent root initiator with hypocotyl cuttings of

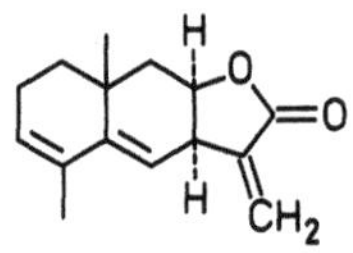

175

Vigna radiata [328]. The sesquiterpene lactones confertiflorin **176** and its desacetyl derivative **177** from *Ambrosia confertiflora*, 11,13-dihydroparthenolide **178** from *A. artemisiifolia* and parthenin **179** from *Parthenium hysterophorus* were shown to be witchweed germination stimulants [329]. The activity of compound **178** is comparable to the activity of strigol. The germination stimulation in *Avena fatua* by synthetic strigol analogs was also studied [336].

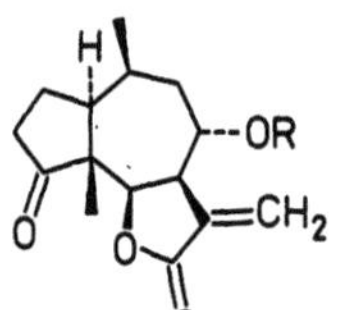

176 R = Ac
177 R = H

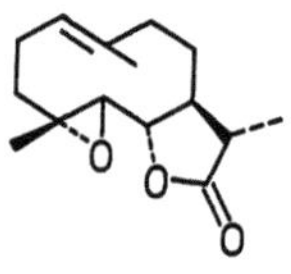

178

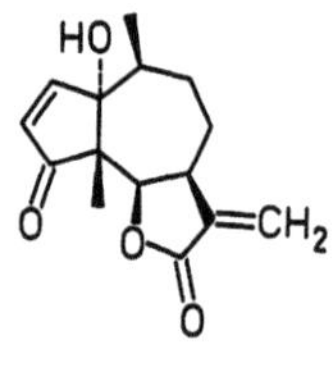

179

From light-grown dwarf pea shoots pisatin **180** and xanthoxin **181** were isolated as endogenous neutral inhibitors [330]. Kaimonolide A and B are novel macrolide plant growth inhibitors which were isolated from the culture medium of a *Strepto-*

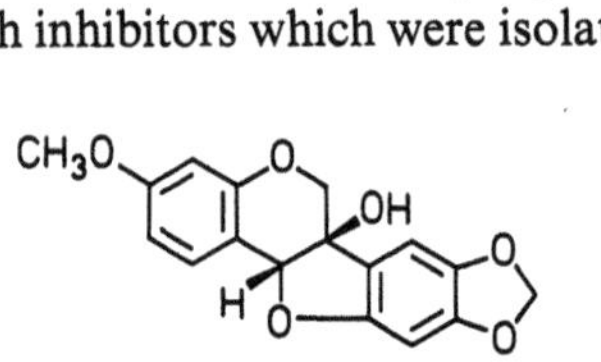

180

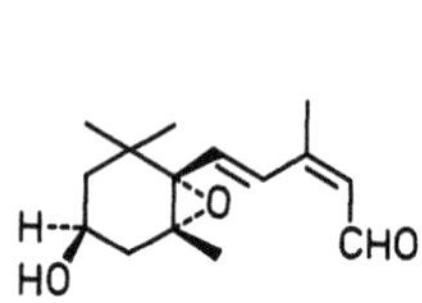

181

myces strain and structurally elucidated [331, 359]. The pathogenic fungus *Rosellinia necatrix* produces rosellichalasin **182** which has been found to exhibit seed germination inhibitory activity [332].

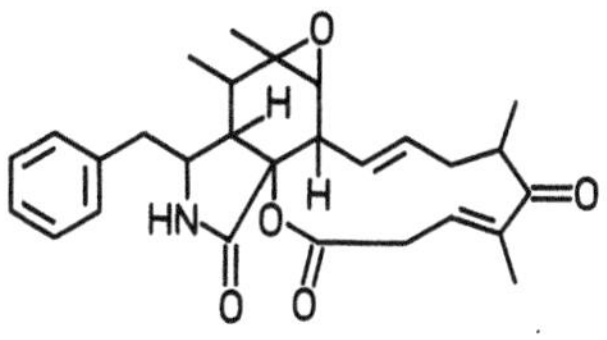

182

Recently 3-(methylthio)-methylene-2-pyrrolidinethione **183** was isolated as growth inhibitor from radish seedlings [337].

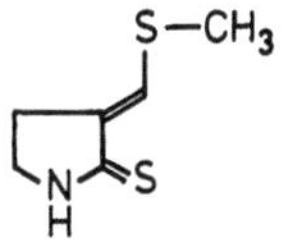

183

The effect of triacontanol **1** on photosynthesis, alkaloid content and growth in *Papaver somniverum* was studied [340]. Triacontanol was found to elicit a new plant growth substance in rice which was identified as 9-β-L(+)-adenosine [341]. The naturally occurring avocado rooting promotor **5** was shown to possess 2R,4R configuration [342].

It was suggested that the prostaglandins of *Populus balsamifera* (cf. 2.3.1) are built up from arachidonic acid [343].

The binding sites of turgorin I **57** were found to be located on plasma membranes from *Mimosa pudica* [344].

The hydroxyproline-containing diketopiperazines **94** and **95** not only act as germination promotors for rice seedlings but also as elicitors of resistance against water stress [346, 347].

The quinolizidine alkaloid, matrine, was reported to exhibit plant growth stimulating activities [348].

A strain of the fungus, *Cochliobolus spicifer*, produces spiciferone A **172** [321] and spicifernin **184** [349]. Both compounds at 3×10^{-5} M promoted rice seedling elongation.

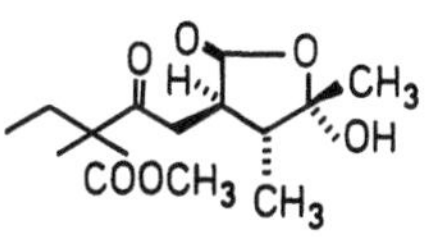

184

The distribution of nicotianamine **120** in different organs of sunflower, tomato and barley has been studied [350, 351]. The effect of nicotianamine **120** on iron remobilisation in de-rooted tomato seedlings has been described [362]. Feeding

(1-^{13}C) L-methionine to barley plants it was shown that three L-methionine units were incorporated into mugineic acid **121** and 2′-deoxymugineic acid **122** [363].

Artemisinin and related sesquiterpene lactones produced in leaves of *Artemisia annua* inhibited seed germination, seedling growth and root induction [352].

5-Methoxy-3-pentadeca-8′,11′,14′-trienylbenzene-1,2,4-triol **185** was isolated from the root exudate of *Sorghum bicolor* [355, 364]. It represents a natural host-derived germination stimulant for the seeds of the parasitic plant *Striga asiatica*

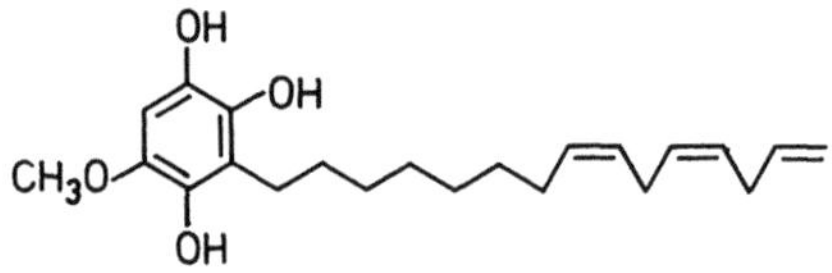

185

which requires this stimulus for germination. Recently, the chemical synthesis of **185** has been achieved [356]. The sesquiterpene strigol **147** also possess germination stimulating activity for *Striga* seeds [364], but is was isolated from cotton as a non-host plant. Now some of its synthetic precursors and a number of natural and synthetic analogs were tested as seed germination stimulants of *Orobanche ramosa* and *Striga asiatica* [353, 354] and *Avena fatua* [336]. Eupatoriochromene and encecalin of *Centaurea solstitialis* were shown to be plant growth regulators [357].

The synthesis of the methyl ester of GA_{73} **160** isolated as a potent antheridiogen from the fern *Lygodium japonicum* was published [358].

Some volatile monoterpenes which are released from leaves of *Thuja occidentalis* inhibited the germination of seeds of *Amaranthus caudatus* and *Lepidium sativum* [360]. Their possible allelopathic effect was discussed.

14 Conclusion

Modern separation and identification techniques such as combined gas chromatography — mass spectrometry, high performance liquid chromatography, immunoassay and other analytic methods, and physico-chemical procedures, have been greatly improved during the last decade. This substantial progress and an extensive screening programme using different plant bioassays has led to the detection, isolation and identification of a high number of natural products produced by microorganisms and by plants. Both have long been recognized as a rich source of biologically active substances. Many of these naturally occurring organic compounds have been shown to elicit regulatory responses in a plant or a plant part and to affect propagation, rooting, flowering, fruiting, dormancy, abscission, senescence, organogenese, growth or other physiological processes. The inhibitory or stimulatory activity has been monitored by various bioassays with different plant species. It seems necessary in the near future to develop and to use internationally accepted standards for plant bioassays in order to compare and evaluate precisely the biological properties of novel plant growth regulating substances of plant or microbial origin.

As shown in this review, and in earlier surveys, endogenous growth-regulating substances belong to quite different chemical groups ranging, for example, from simple aliphatic compounds to structurally complicated terpenes, such as the antheridiogens. This high structural diversity indicates, in all probability, that these biologically active compounds are involved in different mechanisms of plant growth and developmental processes. Most of them affect plant growth development only after exogenous application, but do not have an endogenous regulatory role like the classic groups of phytohormones. However, natural compounds may be important for agrochemical use and as templates for chemical synthesis of novel synthetic plant growth regulators or bioactive compounds with optimized or altered properties. Thus, biologically active, natural products can serve as synthetic models in active substance development, as discussed by Adam [286], Büchel [297] and Schmidt [289].

An area of growing interest is that of naturally occurring plant growth regulatory substances which possess phytohormone-like properties. For example, jasmonic acid, polyamines, metal chelating compounds, brassinosteroids, and oligosaccharins are now under intensive investigation. However, much more research is required to improve understanding of the role in plant growth regulation and of the mode of action of plant and fungal derived compounds which elicit inhibitory or stimulatory responses in plants.

Acknowledgement: Particular thanks are due to Mrs. Monika Kummer for her valuable help and technical assistance in preparing the manuscript. The author wishes to thank Mrs. Johanna Leipnitz and Mrs. Antje Betram for typewriting the manuscript and Mr. E. Schäfer for the accurate drawing of formulas.

15 References

1. Davis PJ (1987) Plant hormones and their role in plant growth and development. Martinus Nijhoff, Dordrecht
2. MacMillan J (ed) (1980) Hormonal regulation of development I. Encyclopedia of plant physiology. New Series vol 9, Springer, Berlin Heidelberg New York
3. Rivier L, Crozier A (1987) Principles and practice of plant hormone analysis. vols 1 and 2, Academic, London
4. Sembdner G, Schneider G, Schreiber K (eds) (1987) Methoden zur Pflanzenhormonanalyse, Fischer, Jena
5. Takahashi N (ed) (1986) Chemistry of plant hormones, CRC Press, Boca Raton FL
6. Doerffling K, Crocoll C (1988) Progr. Bot. 50: 116
7. Sembdner G, Herrmann G, Schliemann W (1989) Progr. Bot. 51: 134
8. Bearder JR (1980) Plant hormones and other growth substances — their background, structures and occurrence. In: MacMillan J (ed) Hormonal regulation of development. I. Encyclopedia of plant physiology. New Series vol 9, Springer, Berlin Heidelberg New York, pp 9–112
9. Gross D (1975) Phytochemistry 14: 2105
10. Gross D (1980) Z. Chem. 20: 397
11. Letham DS (1978) Naturally-occurring plant growth regulators other than the principal hormones of higher plants. In: Letham DS, Goodwin PB, Higgins TJV (eds.) Phytohormones

and related compounds: A comprehensive treatise. vol I. Elsevier/North Holland Biomedical Press, Amsterdam New York, pp 349–417
12. Mandava NM (1979) Natural products in plant growth regulation. In: Mandava NB (ed) Plant growth substances ACS Symp. Ser. 111: 135–213
13. McLaren JS (1986) Pestic. Sci. 17: 559
14. Schreiber K (1975) Environ. Qual. Saf. Suppl. 3: 483
15. Sembdner G, Gross D (1986) Plant growth substances of plant and microbial origin. In: Bopp M (ed) Plant growth substances. Springer, Berlin Heidelberg New York, pp 139–146
16. Takahashi N (1986) Introduction. In: Takahashi N (ed) Chemistry of plant hormones. CRC Press, Boca Raton FL, pp 1–7
17. Charlton JL, Hunter NR, Green NA, Fritz W, Addison BM, Woodbury W (1980) Can. J. Plant Sci. 60: 795
18. Lewak S, Skowronska E (1982) Physiol. Plant. 55: 183
19. Hoagland RE (1980) Bot. Gaz. (Chicago) 141: 53
20. Houtz RL, Ries SK, Tolbert NE (1985) Plant Physiol. 79: 357
21. Lesniak AP, Haug A, Ries SK (1986) Physiol. Plant. 68: 20
22. Mamat ASB, Fontenot JF, Newsom DW (1983) Hort. Sci. 18: 247
23. Eriksen AB, Sellden G, Skogen D, Nilsen S (1981) Planta 152: 44
24. Henry EW, Gordon CJ (1980) J. Exp. Bot. 31: 1297
25. Ramani S, Kannan S (1980) Z. Pflanzenphysiol. 99: 427
26. Ries S (1985) CRC Crit. Rev. Plant Sci. 2: 239
27. Ries SK, Houtz R (1983) Hort Sci. 18: 654
28. Ries SK, Wert V (1982) J. Plant Growth Regul. 1: 117
29. Knowles NR, Ries SK (1981) Plant Physiol. 68: 1279
30. Iyer R, Mamdapur VR (1986) Chem. Ind.: 398
31. Qian Ch, Deng D, Qiao Y, Zheng Z, Chen J, Ye X (1981) Zhiwu Xuebao 23: 423
32. Villemin D (1985) Chem. Ind.: 763
33. Ries SK, Wert VF (1988) Planta 173: 79
34. Raviv M, Reuveni O (1984) J. Am. Soc. Hortic. Sci. 109: 284
35. Raviv M, Becker D, Sahali Y (1986) Plant Growth Regul. 4: 371
36. Raviv M, Reuveni O, Goldschmidt EE (1986) Plant Growth Regul. 4: 95
37. Bittner S, Gazit S, Blumenfeld A (1971) Phytochemistry 10: 1417
38. Ueda J, Kato J (1982) J. Plant Growth Regul. 1: 195
39. Ohkawa M, Nishikawa Y (1987) Plant Sci. 53: 35
40. Ando T, Tsukamoto Y (1981) Phytochemistry 20: 2143
41. Yamane H, Sato Y, Takahashi N, Takeno K, Furuya M (1980) Agric. Biol. Chem. 44: 1697
42. Nair MG, Burke BA (1988) Phytochemistry 27: 3169
43. Matsusaki T, Koseki K, Koiwai A (1988) Agric. Biol. Chem. 52: 1889
44. Nishikawa Y, Yoshimoto K, Nishijima M, Fukuoka F, Ikekawa T (1981) Chem. Pharm. Bull. 29: 505
45. Matsuzaki T, Miyano M, Yasumatsu N, Matsushita H, Koiwai A (1988) Agric. Biol. Chem. 52: 1899
46. Oster U, Blos I, Ruediger W (1987) Z. Naturforsch. Teil C 42: 1179
47. Isogai A, Washizu M, Kondo K, Murakoshi S, Suzuki A (1984) Agric. Biol. Chem. 48: 2607
48. Lohaus E, Blos I, Schaefer W, Ruediger W (1982) Z. Naturforsch. Teil C 37: 802
49. Gregson PP, Marwood JF, Quinn RJ (1979) Tetrahedron Lett.: 4505
50. Janistin B (1982) Planta 154: 485
51. Levin ED, Cherepanova VE, Zimovtseva IA, Sedlova TO (1988) Phytochemistry 27: 3241
52. Demole E, Lederer E, Mercier D (1962) Helv. Chim. Acta 45: 675
53. Aldridge DC, Galt S, Giles D, Turner WB (1971) J. Chem. Soc. C.: 1623
54. Cross BE, Webster GRB (1970) J. Chem. Soc. C: 1839
55. Ueda J, Kato J (1980) Plant Physiol. 66: 246
56. Dathe W, Roensch H, Preiss A, Schade W, Sembdner G, Schreiber K (1981) Planta 153: 530
57. Yamane H, Takagi H, Abe H, Yokota T, Takahashi N (1981) Plant Cell Physiol. 22: 689
58. Yamane H, Abe H, Takahashi N (1982) Plant Cell Physiol. 23: 1125

59. Meyer A, Miersch O, Buettner C, Dathe W, Sembdner G (1984) J. Plant Growth Regul. 3: 1
60. Sembdner G, Klose C (1985) Biol. Rundsch. 23: 29
61. Tsurumi S, Asahi (1985) Physiol. Plant. 64: 207
62. Ueda J, Kato J (1981) Z. Pflanzenphysiol. 103: 357
63. Ueda J, Kato J, Yamane H, Takahashi N (1981) Physiol. Plant. 52: 305
64. Satler SO, Thimann KV (1981) C. R. Acad. Sci., Ser. III 293: 735
65. Ueda J, Kato J (1982) Physiol. Plant. 54: 249
66. Yamane H, Sugawara J, Suzuki Y, Shimamura E, Takahashi N (1980) Agric. Biol. Chem. 44: 2857
67. Yamane H, Takahashi N, Ueda J, Kato J (1981) Agric. Biol. Chem. 45: 1709
68. Sembdner G, Dathe W, Bergner C, Roensch H (1983) Wiss. Z. E. M. Arndt-Univ. Greifswald 32: 113
69. Miersch O, Sembdner G (1985) Acta Univ. Agri. (Brno) 3: 101
70. Meyer A, Miersch O, Vorkefeld S, Sembdner G (1985) Acta Univ. Agr. (Brno) 3: 471
71. Bohlmann F, Wegner P, Jakupovic J, King RM (1984) Tetrahedron 40: 2537
72. Brueckner C, Kramell R, Schneider G, Knoefel H-D, Sembdner G, Schreiber K (1986) Phytochemistry 25: 2236
73. Brueckner C, Kramell R, Schneider G, Schmidt J, Preiss A, Sembdner G, Schreiber K (1988) Phytochemistry 27: 275
74. Herrmann G, Kramell H-M, Kramell R, Weidhase R, Sembdner G (1987) Biological activity of jasmonic acid conjugates. In: Schreiber K, Schuette H.-R, Sembdner G (eds) Conjugated plant hormones, structure, metabolism and function, VEB Deutscher Verlag der Wissenschaften, Berlin, p 315
75. Miersch O, Preiss A, Sembdner G, Schreiber K (1987) Phytochemistry 26: 1037
76. Nishida R, Acree TE (1984) J. Agric. Food Chem. 32: 1001
77. Nishida R, Acree TE, Fukami H (1985) Agric. Biol. Chem. 49: 769
78. Miersch O, Meyer A, Vorkefeld S, Sembdner G (1986) J. Plant Growth Regul. 5: 91
79. Miersch O, Sembdner G, Schreiber K (1989) Phytochemistry 28: 339
80. Fukui H, Koshimizu K, Usuda S, Yamazaki Y (1977) Agric. Biol. Chem. 41: 175
81. Fukui H, Koshimizu K, Yamazaki Y, Usuda S (1977) Agric. Biol. Chem. 41: 189
82. Dathe W, Miersch O, Schmidt J (1989) Biochem. Physiol. Pflanz. 185: 83
83. Kataoka H, Yamada T, Goto K, Tsuji J (1987) Tetrahedron 43: 4107
84. Knoefel H-D, Gross D (1988) Z. Naturforsch. Teil C 43: 29
85. Unverricht A, Gross D (1986) J. Labelled Compds. Radiopharm. 23: 515
86. Kitahara T, Miura K, Warita Y, Takagi Y, Mori K (1987) Agric. Biol. Chem. 51: 1129
87. Vick BA, Zimmerman DC (1983) Biochem. Biophys. Res. Commun. 111: 470
88. Vick BA, Zimmerman DC (1984) Plant Physiol. 75: 458
89. Vick BA, Zimmerman DC (1986) Plant Physiol. 80: 202
90. Vick BA, Zimmerman DC (1987) Plant Physiol. 85: 1073
91. Meyer A, Gross D, Vorkefeld S, Kummer M, Schmidt J, Sembdner G, Schreiber K (1989) Phytochemistry 28: 1007
92. Yoshihara T, Omer EA, Shibata H, Koshino H, Sakamura S, Koda Y, Kikuta Y, Okazawa Y (1988) Chemical structure of a potato tuber-inducing substance. Abstract, 13th Intern. Conf. on Plant Growth Substances, 17–22 July 1988, Calgary, p 427
93. Koda Y, Yoshihara T, Sakamura S, Kikuta Y (1988) Comparisons of biological activities of the potato tuber-inducing substance, jasmonic acid and its derivatives. Abstract 13th Intern. Conf. Plant Growth Substance, 17–22 July 1988, Calgary, p 404
94. Saniewski M, Czapski J, Nowacki J (1987) Biologia Plantarum 29: 17
95. Saniewski M, Nowacki J, Czapski J (1987) J. Plant Physiol. 129: 175
96. Saniewski M, Urbanek H, Czapski J (1987) J. Plant Physiol. 127: 177
97. Saniewski M, Czapski J, Nowacki J, Lange E (1987) Biologia Plantarum 29: 199
98. Saniewski M, Nowacki J, Lange E, Czapski J (1986) Fruit Sci. Rep. 13: 193
99. Weidhase RA, Kramell H-M, Lehmann J, Liebisch HW, Lerbs W, Parthier B (1987) Plant Sci. 51: 177
100. Weidhase RA, Lehmann J, Kramell H-M, Sembdner G, Parthier B (1987) Physiol. Plant. 69: 161

101. Mueller-Uri F, Parthier B, Nover L (1988) Planta 176: 241
102. Herrmann G, Lehmann J, Peterson A, Sembdner G, Weidhase RA, Parthier B (1988) J. Plant Physiol. 134: 703
103. Anderson JM (1988) J. Plant Growth Regul. 7: 203
104. Nakajima H, Hamasaki T, Shikata Y, Kimura Y (1986) Agric. Biol. Chem. 50: 753
105. Isogai A, Sakuda S, Nakayama J, Watanabe S, Suzuki A (1987) Agric. Biol. Chem. 51: 2277
106. Eussen JHH, Niemann GJ (1981) Z. Pflanzenphysiol. 102: 263
107. Kaihara S, Watanabe K, Takimoto A (1981) Plant Cell Physiol. 22: 819
108. Kobayashi S, Ariga M, Ozawa T, Imagawa H (1984) Agric. Biol. Chem. 48: 389
109. Kobayashi S, Ozawa T, Imagawa H (1981) Agric. Biol. Chem. 45: 1519
110. Raskin I, Ehmann H, Melander WR, Meeuse BJD (1987) Science 237: 1601
111. Adachi T, Inagaki K (1988) Agric. Biol. Chem. 52: 2625
112. Chiji H, Tanaka S, Izawa M (1980) Agric. Biol. Chem. 44: 205
113. Itokaw H, Oshida Y, Ikuta A, Inatomi H, Adachi T (1982) Phytochemistry 21: 1935
114. Nishitoba Y, Nishimura H, Nishiyama T, Mizutani J (1987) Phytochemistry 26: 3181
115. Fujioka S, Yamaguchi I, Murofushi N, Takahashi N, Kaihara S, Takimoto A (1983) Plant Cell Physiol. 24: 235
116. Wightman F, Lighty DL (1982) Physiol. Plant. 55: 17
117. Voblikova VD, Kobrina NS, Gerasimova NM, Pavlova ZN, Dem'yanova GF, Murygina VP, Volosova LI, Muromtsev GS (1985) Khim. Prir. Soedin: 387
118. Kobayashi S, Ozawa T, Imagawa H (1982) Agric. Biol. Chem. 46: 845
119. Suzuki Y, Yamaguchi I, Murofushi N, Takahashi N (1988) Agric. Biol. Chem. 52: 1013
120. Crow WD, Osawa T, Paton DM, Willing RR (1977) Tetrahedron Lett.: 1073
121. Bolte ML, Bowers J, Crow WD, Paton DM, Sakurai A, Takahashi N, Ujiie M, Yoshida S (1984) Agric. Biol. Chem. 48: 373
122. Bolte ML, Crow WD, Yoshida S (1982) Aust. J. Chem. 35: 1411
123. Bolte ML, Crow WD, Takahashi N, Sakurai A, Ujiie M, Yoshida S (1985) Agric. Biol. Chem. 49: 76
124. Yoneyama K, Konnai M, Takematsu T, Yoshida S, Takahashi N, Crow WD (1987) Agric. Biol. Chem. 51: 2607
125. Lee TS, Purse JG, Pryce RJ, Horgan R, Wareing PF (1981) Planta 152: 571
126. Hasegawa K, Hase T (1981) Plant Cell Physiol. 22: 303
127. Hasegawa K, Miyamoto K (1980) Plant Cell Physiol. 21: 363
128. Hase T, Hasegawa K (1982) Phytochemistry 21: 1021
129. MacRae WD, Towers GHN (1984) Phytochemistry 23: 1207
130. Schildknecht H (1983) Angew. Chem. 95: 689
131. Schildknecht H (1984) Endeavour 8 (3): 113
132. Schildknecht H (1986) Spektr. Wiss. 11: 44
133. Schildknecht H, Schumacher K (1982) Pure Appl. Chem. 54: 2501
134. Schildknecht H, Schumacher K (1981) Chem. Zeitung 105: 287
135. Schildknecht H, Schumacher K (1982) Chem. Zeitung 106: 421
136. Schildknecht H, Bender W (1983) Chem. Zeitung 107: 111
137. Paterson NW, Weyers JDB, Schildknecht H (1987) J. Plant Physiol. 128: 491
138. Schildknecht H, Sauer H, Kunzelmann P (1988) Z. Naturforsch. Teil C 43: 519
139. Schildknecht H, Kresbach GM, Muley R (1984) Chem. Zeitung 108: 399
140. Schildknecht H, Milde H (1987) Carbohydr. Res. 164: 23
141. Schildknecht H, Muley R, Kresbach GM, Kunzelmann P, Krauss D (1986) Z. Naturforsch. Teil C 41: 547
142. Schildknecht H, Schumacher K, Soellner R (1983) Chem. Zeitung 107: 233
143. Inamori Y, Kato Y, Kubo M, Yasuda M, Baba K, Kozawa M (1984) Chem. Pharm. Bull. 32: 213
144. Hashimoto T, Tori M, Asakawa Y (1988) Phytochemistry 27: 109
145. Watanabe K, Miyakado M, Iwai T, Izumi K, Yanagi K (1988) Agric. Biol. Chem. 52: 1079
146. Rietveld WJ (1983) J. Chem. Ecol. 9: 295
147. Kimura Y, Shimada A, Nakajima H, Hamasaki T (1988) Agric. Biol. Chem. 52: 1253

148. Kobayashi A, Yata S, Hino T, Kawazu K (1987) Agric. Biol. Chem. 51: 2857
149. Jansson E, Svensson S-B (1980) Physiol. Plant. 48: 486
150. Shimomura H, Sashida Y, Nakata H, Kawasaki J, Ito Y (1982) Phytochemistry 21: 2213
151. Ohigashi, H, Minami S, Fukui H, Koshimizu K, Mizutani F, Sugiura A, Tomana T (1982) Agric. Biol. Chem. 46: 2555
152. Spencer GF, Tjarks LW (1985) J. Plant Growth Regul. 4: 177
153. Evans LS, Tramontano WA, Gill R (1987) Phytochemistry 26: 2891
154. Lynn DG, Jaffe K, Cornwall M, Tramontano W (1987) J. Am. Chem. Soc. 109: 5858
155. Cardellina JH, Nigh D, VanWagenen BC (1986) J. Nat. Prod. 49: 1065
156. Yomosa K, Hirota A, Sakai H, Isogai A (1987) Agric. Biol. Chem. 51: 921
157. Raub MF, Cardellina II JH, Schwede JG (1987) Phytochemistry 26: 619
158. Lebrun MH, Nicolas L, Boutar M, Gaudemer F, Ranomenjanahary S, Gaudemer A (1988) Phytochemistry 27: 77
159. Morris CF, Mueller DD, Faubion JM, Paulsen GM (1988) Plant Physiol. 88: 435
160. Mandava N, Anderson JD, Dutky SR (1974) Phytochemistry 13: 2853
161. Ienaga K, Nakamura K, Goto T, Konishi J (1987) Tetrahedron Lett. 28: 4587
162. Ienaga K, Nakamura K, Goto T (1987) Tetrahedron Lett. 28: 1285
163. Cutler HG, Springer JP, Arrendale RF, Arison BH, Cole PD, Roberts RG (1988) Agric. Biol. Chem. 52: 1725
164. Kusumi T, Chang CG, Wheeler M, Kubo I, Nakanishi K, Naoki H (1981) Tetrahedron Lett. 22: 3451
165. Hase T, Koreeda M, Hasegawa K (1983) Phytochemistry 22: 1275
166. Hasegawa K, Shiihara S, Iwagawa T, Hase T (1982) Plant Physiol. 70: 626
167. Niemeyer HM (1988) Phytochemistry 27: 3349
168. Wolf RB, Spencer GF, Plattner RD (1985) J. Nat. Prod. 48: 59
169. Barnes JP, Putnam AR, Burke BA, Aasen AJ (1987) Phytochemistry 26: 1385
170. Aiupova AT, Molchanov LV, Kadryov CS, Aliev NA, Giasov K, Loi NP, Tsoi Z, Umarov AA (1979) Agrokhimiia 10: 107
171. Nakanishi K, Patt LS, Occolowitz JL, Almeida MS, Evans LS (1978) J. Am. Chem. Soc. 100: 7759
172. Evans LS (1979) Am. J. Bot. 66: 880
173. Evans LS, Almeida MS, Lynn DG, Nakanishi K (1979) Science 203: 1122
174. Evans LS, Tramontano WA (1981) Am. J. Bot. 68: 1282
175. Tramontano WA, Hartnett CM, Lynn DG, Evans LS (1982) Phytochemistry 21: 1201
176. Tramontano W, Lynn DG, Evans LS (1983) Phytochemistry 22: 343
177. Lynn DG, Lewis DH, Tramontano WA, Evans LS (1984) Phytochemistry 23: 1225
178. Wink M (1983) Planta 158: 365
179. Mandava NB, Worley JF, Kapadia GJ (1981) J. Nat. Prod. 44: 94
180. Hamasaki T, Nakajima H, Yokota T, Kimura Y (1983) Agric. Biol. Chem. 47: 891
181. Cutler HG, Cox RH, Crumley FG, Cole PD (1986) Agric. Biol. Chem. 50: 2943
182. Kimura Y, Hamasaki T, Nakajima H (1986) Agric. Biol. Chem. 50: 1649
183. Teijin Ltd. (1985) Jpn. Kokai Tokkyo Koho JP 6001, 176 [8501, 176]. (1985) Chem. Abstr. 103, 19769
184. Lohaus E, Zenger C, Ruediger W, Cmiel E (1985) Z. Naturforsch. Teil C 40: 490
185. Dhawan AK, Paton DM, Willing RR (1979) Planta 146: 419
186. Kalsi PS, Singh J, Crow WD, Chhabra BR (1987) Phytochemistry 26: 3367
187. Nakajima H, Nishimura K, Hamasaki T, Kimura Y, Yokota T, Udagawa S (1987) Agric. Biol. Chem. 51: 1221
188. Nakajima H, Nishimura K, Hamasaki T, Kimura Y, Udagawa S (1987) Agric. Biol. Chem. 51: 2831
189. Noma M, Noguchi M, Tamaki E (1971) Tetrahedron Lett.: 2017
190. Kristensen I, Larsen PW (1974) Phytochemistry 13: 2791
191. Noma M, Noguchi M (1976) Phytochemistry 15: 1701
192. Budesinsky M, Budzikiewicz H, Prochazka Z, Ripperger H, Roemer A, Scholz G, Schreiber K (1980) Phytochemistry 19: 2295
193. Ripperger H, Schreiber K (1982) Heterocycles 17: 447

194. Prochazka Z, Scholz G (1984) Experientia 40: 794
195. Schreiber K (1986) Pure Appl. Chem. 58: 745
196. Scholz G, Becker R, Stephan UW, Rudolph A, Pich A (1988) Biochem. Physiol. Pflanz. 183: 257
197. Benes I, Schreiber K, Ripperger H, Kircheiss A (1983) Experientia 39: 261
198. Scholz G, Faust J, Ripperger H, Schreiber K (1988) Phytochemistry 27: 2749
199. Takemoto T, Nomoto K, Fushiya S, Ouchi R, Kusano G, Hikino H, Takagi S, Matsuura Y, Kakudo M (1978) Proc. Jpn Acad. Ser. B 54: 469
200. Ohfune Y, Tomita M, Nomoto K (1981) J. Am. Chem. Soc. 103: 2409
201. Nomoto K, Yoshioka H, Arima M, Fushiya S, Takagi S, Takemoto T (1981) Chimia 35: 249
202. Nomoto K, Yoshioka H, Takemoto T, Fushiya S, Nozoe S, Takagi S (1979) 22. Symposium Chem. Natural Products, Symposium Papers, Fukuoka, p 619
203. Fushiya S, Sato Y, Nozoe S, Nomoto K, Takemoto T, Takagi S (1980) Tetrahedron Lett. 21: 3071
204. Fushiya S, Sato Y, Nozoe S (1980) Chem. Lett.: 1215
205. Sugiura Y, Tanaka H, Mino Y, Ishida T, Ota N, Inoue M, Nomoto K, Yoshioka H, Takemoto T (1981) J. Am. Chem. Soc. 103: 6979
206. Sugiura Y, Nomoto K (1984) Struct. Bonding 58: 107
207. Mino Y, Ishida T, Ota N, Inoue M, Nomoto K, Takemoto T, Tanaka H, Sugiura Y (1983) J. Am. Chem. Soc. 105: 4671
208. Neilands JB, Leong SA (1986) Annu. Rev. Plant Physiol. 37: 187
209. Leong J (1986) Annu. Rev. Phytopathol. 24: 187
210. Grill E, Winnacker EL, Zenk MH (1985) Science 230: 674
211. Grill E, Zenk MH, Winnacker E-L (1985) Naturwissenschaften 72: 432
212. Grill E, Winnacker EL, Zenk MH (1987) Proc. Nat. Acad. Sci. 84: 439
213. Grill E, Winnacker E-L, Zenk MH (1988) Experientia 44: 539
214. Gekeler W, Grill E, Winnacker E-L, Zenk MH (1988) Arch. Microbiol. 150: 197
215. Heby O (1981) Differentiation 19: 1
216. Pegg AE (1986) Biochem. J. 234: 249
217. McCann PP, Pegg AE, Sjoerdsma A (1987) Inhibition of polyamine metabolism. Academic, San Diego
218. Slocum RD, Kaur-Sawhney R, Galston AW (1984) Arch. Biochem. Biophys. 235: 283
219. Tabor CW, Tabor H (1984) Annu. Rev. Biochem. 53: 749
220. Tabor CW, Tabor H (1985) Microbiol. Rev. 49: 81
221. Smith TA (1985) Annu. Rev. Plant Physiol. 36: 117
222. Serafini-Fracassini D (1984) Effects of polyamines and polyamine inhibitors on growth, differentiation and organogenesis in plant tissue culture. In: Caldarera CM, Bachrach U (eds) Advances in polyamines in biomedical science, CLUEB Bologna. pp 189–196
223. Burtin D, Martin-Tanguy J, Paynot M, Rossin N (1989) Plant Physiol. 89: 104
224. Christ M, Felix H, Harr J (1989) Z. Naturforsch. Teil C 44: 49
225. Felix H, Harr J (1989) Z. Naturforsch., Teil C 44: 55
226. Masse J, Laberche JC, Jeanty G (1988) Plant Growth Regul. 7: 249
227. Kaur-Sawhney R, Tiburcio AF, Galston AW (1988) Planta 173: 282
228. Tiburcio AF, Kaur-Sawhney R, Galston AW (1988) Plant Cell Physiol. 29: 1241
229. Torrigiani P, Altamura MM, Pasqua G, Monacelli B, Serafini-Fracassini D, Bagni N (1987) Physiol. Plant. 70: 453
230. Komai K, Iwamura J, Hamada M, Ueki K (1986) Zasso Kenkyu 31: 280; Chem. Abstr. (1987) 106: 192933
231. Cameron DW, Feutrill GJ, Perlmutter P, Sasse JM (1984) Phytochemistry 23: 533
232. Shimomura H, Sashida Y, Nakata H, Yamamoto A, Kawakuba Y, Kawasaki J (1983) Plant Cell. Physiol. 24: 123
233. Schliemann W, Adam G (1982) Phytochemistry 21: 1438
234. Oritani T, Ichimura M, Yamashita K (1984) Agric. Biol. Chem. 48: 1677
235. Oritani T, Niitsu M, Kato T, Yamashita K (1985) Agric. Biol. Chem. 49: 2819
236. Sassa T, Mitobe H, Haruki E (1988) Agric. Biol. Chem. 52: 1625

237. Matsuo A, Kubota N, Uto S, Nakayama M, Hayashi S, Yamasaki K (1979) Chem. Lett.: 1383
238. Matsuo A, Uto S, Nozaki H, Nakayama M, Hayashi S (1980) J. Chem. Soc. Chem. Comm.: 1220
239. Matsuo A, Uto S, Nozaki H, Nakayama M (1984) J. Chem. Soc. Perkin Trans I: 215
240. Kodama M, Tambunan USF, Tsunoda T, Ito S (1986) Bull. Chem. Soc. Jpn. 59: 1897
241. Matsuo A, Atsumi K, Nadaya K, Nakayama M, Hayashi S (1981) Phytochemistry 20: 1065
242. Matsuo A, Nadaya K, Nakayama M, Hayashi S (1981) Nippon Kagaku Kaishi: 665
243. Watanabe S, Kobayashi A, Yamashita K (1981) Agric. Biol. Chem. 45: 2919
244. Cutler HG, Cole PD, Arrendale RF, Bassfield RL, Cox RH, Roberts RG, Hanlin RT (1986) Agric. Biol. Chem. 50: 2667
245. Kalsi PS, Goyal R, Talwar KK, Chhabra BR (1988) Phytochemistry 27: 2079
246. Spring O, Albert K, Gradmann W (1981) Phytochemistry 20: 1883
247. Spring O, Hager A (1982) Planta 156: 433
248. Johnson AW, Gowda G, Hassanali A, Knox J, Monaco S, Razavi Z, Rosebery G (1981) J. Chem. Soc., Perkin I: 1734
249. Hsiao AJ, Worsham AD, Moreland DE (1981) Z. Pflanzenphysiol. 104: 1
250. Kalsi PS, Talwar KK (1981) Phytochemistry 20: 511
251. Corey EJ, Myers AG, Takahashi N, Yamane H, Schraudolf H (1986) Tetrahedron Lett. 27: 5083
252. Corey EJ, Myers AG (1985) J. Am. Chem. Soc. 107: 5574
253. Furber M, Mander LN (1987) J. Am. Chem. Soc. 109: 6389
254. Yamane H, Nohara K, Takahashi N, Schraudolf H (1987) Plant Cell Physiol. 28: 1203
255. Yamane H, Satoh Y, Nohara K, Nakayama M, Murofushi N, Takahashi N, Takeno K, Furuya M, Furber M, Mander LN (1988) Tetrahedron Lett. 29: 3959
256. Yamane H, Takahashi N, Takeno K, Furuya M (1979) Planta 147: 251
257. Nester JE, Schedlbauer MD (1982) Can. J. Bot. 60: 1606
258. Nester JE, Coolbaugh RC (1986) Plant Physiol. 82: 230
259. Nester JE, Veysey S, Coolbaugh RC (1987) Planta 170: 26
260. Furber M, Mander LM, Nester JE, Takahashi N, Yamane H (1989) Phytochemistry 28: 63
261. Furber M, Mander LN (1988) J. Am. Chem. Soc. 110: 4084
262. Furber M, Mander LN (1988) Tetrahedron Lett. 29: 3339
263. Kalsi PS, Gupta D, Dhillon RS, Arora GS, Talwar KK, Wadia MS (1981) Phytochemistry 20: 1539
264. Kalsi PS, Sood VB, Masih AB, Gupta D, Talwar KK (1983) Phytochemistry 22: 1387
265. Kalsi PS, Kaur G, Sharma S, Talwar KK (1984) Phytochemistry 23: 2855
266. Sasse JM, Rowan KS, Galbraith MN (1981) Phytochemistry 20: 2195
267. Sasse JM, Wardrop JJ, Rowan KS, Aspinall D, Coombe BG, Paleg LG, Buta JG (1982) Physiol. Plant. 55: 51
268. Hoffmann HMR, Rabe J (1985) Angew. Chem. 97: 96
269. Geuns, JMC (1981) Monogr.-Br. Plant Growth Regul. Group 6: 209
270. Roy S, Dutta AK, Chakraborty DP (1982) Phytochemistry 21: 2417
271. Jarvis BC, Booth A (1981) Physiol. Plant. 53: 213
272. Jarvis BC, Yasmin S, Coleman MT (1985) Physiol. Plant. 64: 53
273. Buchala AJ, Schmid A (1979) Nature 280: 230
274. Buchala AJ, Pythoud F (1988) Physiol. Plant. 74: 391
275. Schmid A, Buchala AJ (1987) J. Plant Growth Regul. 5: 175
276. Albersheim P, Darvill AG (1985) Sci. Amer. 253: 44
277. Dixon RA (1986) Biol. Rev. 61: 239
278. Ebel J (1986) Annu. Rev. Plant Phytopathol. 24: 235
279. Ryan CA (1987) Annu. Rev. Cell Biol. 3: 295
280. Ryan CA (1988) Biochemistry 27: 8879
281. Sharp JK, McNeil M, Albersheim P (1984) J. Biol. Chem. 259: 11321
282. Darvill AG, Davis KR, Doares SH, Gollin DJ, O'Neil RA, Toubart PR, York WS, Albersheim P (1987) Modern Cell Biol. 5: 283

283. Tran Thanh Van K, Toubart P, Cousson A, Darvill AG, Gollin DJ, Chelf P, Albersheim P (1985) Nature 314: 615
284. Sakurai N, Inouhe M, Masuda Y, Kuraishi S (1983) Plant Cell Physiol. 24: 317
285. Berlage U, Schmidt J, Peters U, Welzel P (1987) Tetrahedron Lett. 28: 3091
286. Adam G (1981) Z. Chem. 21: 273
287. Buechel KH (1979) Naturwissenschaften 66: 173
288. Pythoud F, Buchala AJ, Schmid A (1986) Physiol. Plant. 68: 93
289. Schmidt HG (1981) Umschau 81: 25
290. Fujioka S, Yamaguchi I, Murofushi N, Takahashi N, Kaihara S, Takimoto A (1983) Plant Cell Physiol. 24: 241
291. Fujioka S, Yamaguchi I, Murofushi N , Takahashi N, Kaihara S, Takimoto A, Cleland CF (1985) Plant Cell Physiol. 26: 655
292. Fujioka S, Sakurai A, Yamaguchi I, Murofushi N, Takahashi N, Kaihara S, Takimoto A, Cleland CF (1986) Plant Cell Physiol. 27: 1297
293. Fujioka S, Yamaguchi I, Murofushi N, Takahashi N, Kaihara S, Takimoto A, Cleland CF (1986) Plant Cell Physiol. 27: 103
294. Fujioka S, Yamaguchi I, Murofushi N, Takahashi N, Kaihara S, Takimoto A, Cleland CF (1986) Plant Cell Physiol. 27: 109
295. Fujioka S, Sakurai A, Yamaguchi I, Murofushi N, Takakhashi N, Kaihara S, Takimoto A (1987) Plant Cell Physiol. 28: 995
296. Kimura Y, Nakajima H, Hamasaki T, Sugawara I, Parkanyi L, Clardy J (1989) Tetrahedron Lett. 30: 1267
297. Haeggquist ML, Widell K-O, Fredriksson M, Liljenberg C (1988) Physiol. Plant 72: 414
298. Haeggquist ML, Strid L, Widell K-O, Liljenberg C (1988) Physiol. Plant 72: 423
299. Evans LS, Tramontano WA, O'Connor RF, Gill R, Ciancaglini E (1984) Envir. Expt. Botany 24: 283
300. Evans LS, Tramontano WA (1984) Phytochemistry 23: 1837
301. Siddiqui KAI, Gupta AK, Paul AK, Banerjee AK (1979) Experientia 35: 1222
302. Dhar TK, Siddiqui KAI, Ali E (1982) Tetrahedron Lett. 23: 5459
303. Springer JP, Cutler HG, Crumley FG, Cox RH, Davis EE, Thean JE (1981) J. Agric. Food Chem. 29: 853
304. Cole RJ, Wells JM, Cox RH, Cutler HG (1981) J. Agric. Food Chem. 29: 205
305. Cole RJ, Wilson DM, Harper JL, Cox RH, Cochran TW, Cutler HG, Bell DK (1982) J. Agric. Food Chem. 30: 301
306. Kamata S, Sakai H, Hirota A. (1983) Agric. Biol. Chem. 47: 2637
307. Levin ED, Cherepanova VE, Sedlova TO, Zimovtseva IA, Zinchenko SV (1988) Chem. Nat. Compd. 24: 293
308. Cardellina II JH, Raub MF, Van Wagenen (1987) ACS Symp. Ser. 330: 562
309. Bradow JM, Connick WJ (1987) J. Chem. Ecol. 13: 185; (1988) J. Chem. Ecol. 14: 1617, 1633
310. Connick WJ, Bradow JM, Legendre MG, Vail SL, Menges RM (1987) J. Chem. Ecol. 13: 463
311. Brand JM, Page HM, Lindner WA, Markovetz, AJ (1989) Naturwissenschaften 76: 277
312. Yoshihara T, Omer ESA, Koshino H, Sakamura S, Kikuta Y, Koda Y (1989) Agric. Biol. Chem. 53: 2835
313. Kubo I (1989) Pure Appl. Chem. 61: 373
314. Greca MD, Monaco P, Previtera L, Aliotta G, Pinto G, Pollio A (1989) Phytochemistry 28: 2319
315. Ueda J, Yokota T, Takahashi N, Yoshita M, Kato J (1986) Agric. Biol. Chem. 50: 3083
316. Ueda J (1989) Physiol. Plant. 76: 42
317. Ikeda M, Tulloch AP, Hoffman LL (1989) Agric. Biol. Chem. 53: 569
318. Wright AE, Schäfer M, Midland S, Munnecke DE, Sims JJ (1989) Tetrahedron Lett. 30: 5699
319. Cline EI, Adesanya SA, Ogundana SK, Roberts MF (1989) Phytochemistry 28: 2621
320. Cultur HG, Arrendale RF, Cole PD, Davis EE, Cox RH (1989) Agric. Biol. Chem. 53: 1975
321. Nakajima H, Hamasaki T, Kimura Y (1989) Agric. Biol. Chem. 53: 2297
322. Ohigashi H, Kaji M, Sakaki M, Koshimizu K (1989) Phytochemistry 28: 1365
323. Anderegg G, Ripperger H (1989) J. Chem. Soc. Chem. Comm.: 647

324. Becker R, Pich A, Scholz G, Seifert K (1989) Physiol. Plant. 76: 47
325. Stephan UW, Grün M (1989) Biochem. Physiol. Pflanzen 185: 189
326. Oida F, Ota N, Mino Y, Nomoto K, Sugiura Y (1989) J. Am. Chem. Soc. 111: 3436
327. Gekeler W, Grill E, Winnacker EL, Zenk MH (1989) Z. Naturforsch. Teil C 44: 361
328. Kalsi PS, Goyal R, Talwar KK, Chhabra BR (1989) Phytochemistry 28: 2093
329. Fischer NH, Weidenhamer JD, Bradow JM (1989) Phytochemistry 28: 2315
330. Noguchi H, Tsurumi S, Takeuchi Y, Hashimoto T (1989) Physiol. Plant. 76: 74
331. Hirota A, Okada H, Kanza T, Nakayama M, Hirota H, Isogai A (1989) Agric. Biol. Chem. 53: 2831
332. Kimura Y, Nakajima H, Hamasaki T (1989) Agric. Biol. Chem. 53: 1699
333. Sassa T, Yoshida N, Haruki E (1989) Agric. Biol. Chem. 53: 3105
334. Flores HE, Protacio CM, Signs M (1989) Recent Adv. Phytochem. 23: 329
335. Evans PT, Malmberg RL (1989) Annu. Rev. Plant Physiol. Plant Mol. Biol. 40: 235
336. Bradow JM, Connick WJ, Pepperman AB, Wartelle LH (1990) Plant Growth Regul. 9: 35
337. Sakoda M, Hase T, Hasegawa K (1990) Phytochemistry 29: 1031
338. Parthier B (1990) Plant Growth Regul. 9: 57
339. Yoshihara T, Omer EA, Shibata H, Koshino H, Matsuura H, Sakamura S, Koda Y, Kikuta Y, Okazawa Y (1988) Yuki Kagobutso Toronkai Koen Yoshishi 30: 158; (1989) Chem. Abstr. 111: 4252
340. Srivastava NK, Sharma S (1990) Plant Growth Regul. 9: 65
341. Ries S, Wert V, O'Leary NFD, Nair M (1990) Plant Growth Regul. 9: 263
342. Becker D, Sahali Y, Raviv M (1990) Phytochemistry 29: 2065
343. Levin ED, Isaeva EV, Cherepanova VE (1990) Phytochemistry 29: 2325
344. Kallas P, Meier-Augenstein W, Schildknecht H (1990) J. Plant Physiol. 136: 225
345. Kallas P, Meier-Augenstein W, Schildknecht H (1989) Naturwiss. Rundschau 42: 309
346. Ienaga K, Kurohashi M, Nakamura K, Nakanishi T, Ichii T, Konishi J (1988) Experientia 44: 356
347. Ienaga K, Nakamura K. Kurohashi M, Nakanishi T, Ichii T (1990) Phytochemistry 29: 35
348. Zhao Z, Li G, Zhen Q, Shi Y, Chen R (1988) Zhiwu Xuebao 30: 614; Chem. Abstr. (1989) 111: 73010
349. Nakajima H, Hamasaki, T, Maeta S, Kimura Y, Takeuchi Y (1990) Phytochemistry 29: 1739
350. Stephan UW, Scholz G (1990) J. Plant Physiol. 136: 631
351. Stephan UW, Scholz G, Rudolph A (1990) Biochem. Physiol. Pflanzen 186: 81
352. Chen PK, Leather GR (1990) J. Chem. Ecol. 16: 1867
353. Fischer NH, Weidenhamer JD, Riopel JL, Quijano L, Menelaou MA (1990) Phytochemistry 29: 2479
354. Vail SL, Dailey OD, Blanchard EJ, Pepperman AB, Riopel JL (1990) J. Plant Growth Regul. 9: 77
355. Chang M, Netzly DH, Butler LG, Lynn DG (1986) J. Am. Chem. Soc. 108: 7858
356. Sargent MV, Wangchareontrakul S (1990) J. Chem. Soc. Perkin Trans. I 1429
357. Merrill GB (1989) J. Chem. Ecol. 15: 2073
358. Kraft-Klaunzer P, Furber M, Mander LN, Twitchin B (1990) Tetrahedron Lett. 31: 6235
359. Hirota A, Okada H, Kanza T, Nakayama M, Isogai A, Hirota H (1989) Tennen Yuki Kagobutsu Toronkai Koen Yoshishu 656; (1990) Agric. Biol. Chem. 54: 2489
360. Oster U, Spraul M, Rüdiger W (1990) Z. Naturforsch. 45c: 835
361. Kubo I, Ochi M, Shibata K, Hanke FJ, Nakatsu T, Tan KS, Taniguchi M, Kamikawa T, Yamagiwa Y, Arizuka M, Wood WF (1990) J. Nat. Prod. 53: 50
362. Scholz G (1989) Biol. Metals 2: 89
363. Kawai S, Itoh K, Takagi S, Iwashita T, Nomoto K (1988) Tetrahedron Lett. 29: 1053
364. Netzly DH, Riopel JL, Ejeta G, Butler LG (1988) Weed Sci. 36: 441
365. Samson E, Frischmuth K, Berlage U, Heinz U, Hobert K, Welzel P (1991) Tetrahedron 47: 1411
366. Raskin I, Turner IM, Melander WR (1989) Proc. Natl. Acad. Sci. USA 86: 2214
367. Sakoda M, Hasegawa K, Ishizuka K (1991) Phytochemistry 30: 57
368. Rauser WE (1990) Annu. Rev. Biochem. 59: 61
369. Keen NT (1990) ACS Symp. Ser. 439: 114

Synthesis and Labelling of Gibberellins

G. Adam, B. Voigt and M. Lischewski
Institute of Plant Biochemistry, Academy of Sciences, O-4050 Halle/Saale Weinberg, FRG

The gibberellins represent an important class of diterpenoid phytohormones of ubiquitous occurrence in higher plants. On the other hand, they are secondary metabolites in fungi, especially in *Gibberella fujikuroi*. Starting with an introduction into the hitherto identified members it is the aim of the article to give a comprehensive overview of the synthesis of gibberellins. Particularly the pathways to rare native compounds starting from biotechnologically available gibberellic acid (gibberellin A_3), the different methods for the preparation of labelled gibberellins as well as recent developments in the synthesis of structural analogs for structure-activity relationships, molecular probes, and as potential plant growth regulators are reviewed. The total synthesis of gibberellins is not included. An up to date survey of the research in the mentioned field covering the literature through 1988 is provided.

Chemistry of Plant Protection 7

Abbreviations

Dibal-H	Diisobutylaluminium hydride
DMF	N,N′-Dimethylformamide
DMSO	Dimethylsulfoxide
LDA	Lithium diisopropylamide
MCPBA	m-Chloro perbenzoic acid
MOMCl	Methoxymethylchloride
MsCl	Methanesulfonyl chloride
NMMNO	N-Methylmorpholine N-oxide
PCC	Pyridinium chlorochromate
PDC	Pyridinium dichromate
TBDMSCl	t-Butyldimethylsilyl chloride
TsCl	p-Toluenesulfonyl chloride
TsOH	p-Toluenesulfonic acid

1 Introduction

The gibberellins originally have been discovered about 50 years ago as secondary metabolites of the rice pathogen *Gibberella fujikuroi* (Saw.) Woll. (*Fusarium moniliforme* Sheld). Later extensive research showed that they represent an important group of endogenous phytohormones of ubiquitous occurrence in the plant kingdom. Interacting with other hormones and native regulators they control growth and development in higher plants exhibiting a multitude of physiological properties among them stimulation of seed germination, vegetative growth and flowering, breaking of dormancy and inhibition of senescence [1, 2].

2 Naturally Occurring Gibberellins

From the chemical point of view gibberellins are tetracyclic diterpenoids which can be derived following a proposal of Rowe [3] from the *ent*-gibberellane skeleton, whereas in the earlier literature the gibbane nomenclature was used. Until now 74 different gibberellins from higher plants or fungi are known named briefly in the trivial nomenclature as GA_1, GA_2 ... etc. [4]. They belong either to the C_{20}-subgroup having the complete diterpenoid skeleton or to the C_{19}-series which has lost the C-20 carbon, with one exception (GA_{11}) under formation of a 19,10-γ-lactone [5, 6, 7]. This function together with a 6β-carboxylic and a $\Delta^{16(17)}$ terminal methylene group are essential structural features for a high biological activity. The hitherto known members and some of their physical data are summarized in Table 1. Gibberellins can occur endogenously also as conjugates, especially as O-glucosides, glucosyl esters, acetyl derivatives and alkyl esters [8].

Most important member is the biologically high active GA_3 (gibberellic acid, *ent*-3α,10,13-trihydroxy-20-*nor*-gibberella-1,16-diene-7,19-dioic acid 19,10-lactone) available biotechnologically by large scale fermentation of *Gibberella fujikuroi*. From the other native compounds only a GA_4/GA_7 mixture and in small extent GA_{13} can be obtained from a different strain of the mentioned fungus and from GA_3 mother liquors, respectively. From higher plants the gibberellins were isolated only in extremly small yields, whereas colorimetric methods, fluorescence spectroscopy, HPLC and, as the methods of choice, GC-MS analysis of the trimethylsilyl derivatives have been used for identification and quantification. For the structural determination MS- and NMR data are of special importance [7, 9]. The complete ^1H and ^{13}C NMR analysis of GA_3 using two-dimensional techniques was published recently [10].

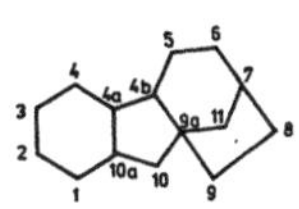

Gibbane

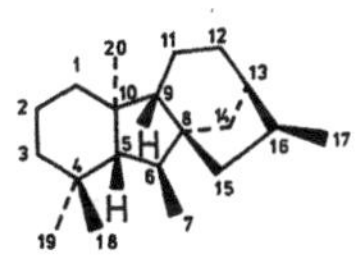

ent-Gibberellane

Table 1. Naturally occurring gibberellins

Compound	Structure	Source F Fungus P Plant	Molecular formula and molecular weight	M. p. in °C	$[\alpha]_D$	References
GA_1	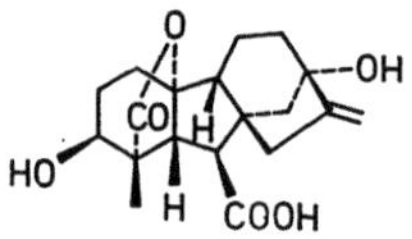	F, P	$C_{19}H_{24}O_6$ 348	255–258	+38°	[11–13]
GA_2	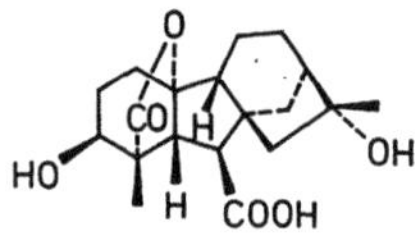	F	$C_{19}H_{26}O_6$ 350	256	+12°	[11, 14]
GA_3	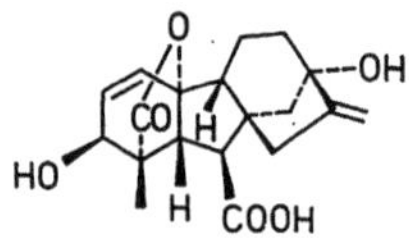	F, P	$C_{19}H_{22}O_6$ 346	234–236	+92°	[15, 16]
GA_4	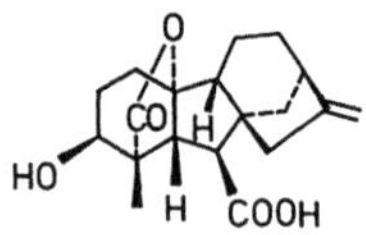	F, P	$C_{19}H_{24}O_5$ 332	222	−15°	[17, 18]
GA_5	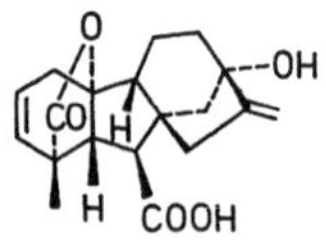	P	$C_{19}H_{22}O_5$ 330	260–261	−77°	[13, 19, 20]
GA_6	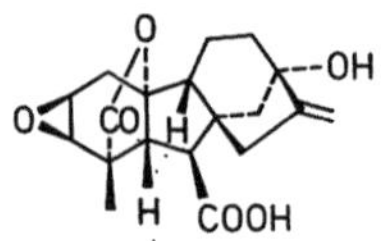	P	$C_{19}H_{22}O_6$ 346	205–209	−28°	[21]
GA_7	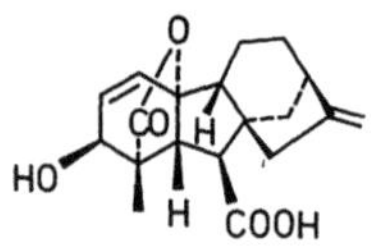	F, P	$C_{19}H_{22}O_5$ 330	169–172 and 203 (dec.)	+20°	[22, 23]
GA_8	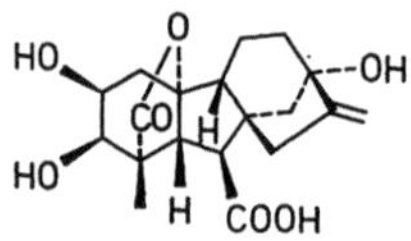	P	$C_{19}H_{24}O_7$ 364	210–215 (dec.)	+30°	[21]
GA_9	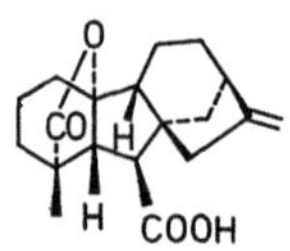	F, P	$C_{19}H_{24}O_4$ 316	208–211	−22°	[23]

Table 1. (Continued)

Compound	Structure	Source F Fungus P Plant	Molecular formula and molecular weight	M. p. in °C	$[\alpha]_D$	References
GA_{10}	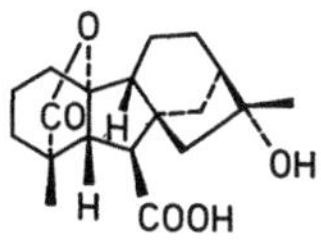	F	$C_{19}H_{26}O_5$ 334	245–246	+3°	[24]
GA_{11}	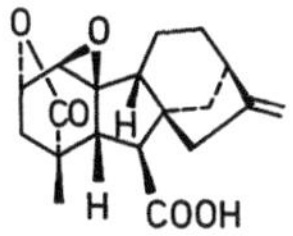	F	$C_{19}H_{22}O_5$ 330	242–244 (dec.)	+11°	[25]
GA_{12}	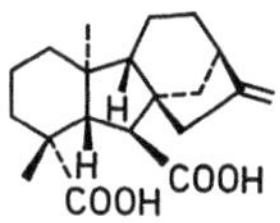	F, P	$C_{20}H_{28}O_4$ 332	245–248		[26]
GA_{13}	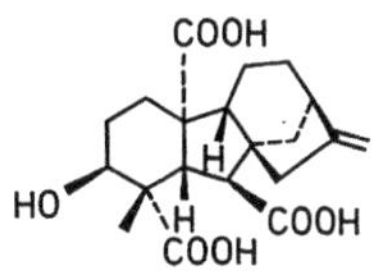	F, P	$C_{20}H_{26}O_7$ 378	194–196 (dec.)	−48°	[27]
GA_{14}	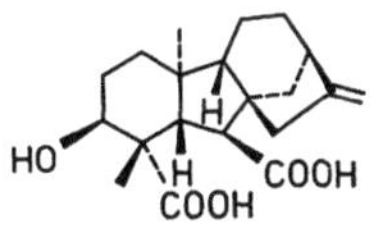	F	$C_{20}H_{28}O_5$ 348	232–233	−73°	[28]
GA_{15}	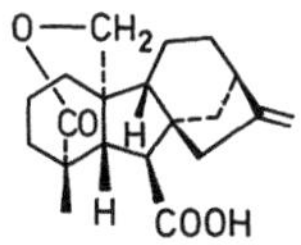	F, P	$C_{20}H_{26}O_4$ 330	274–276	+5°	[29]
GA_{16}	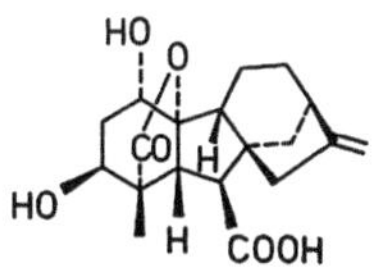	F, P	$C_{19}H_{24}O_6$ 348	137–140		[30, 31]
GA_{17}	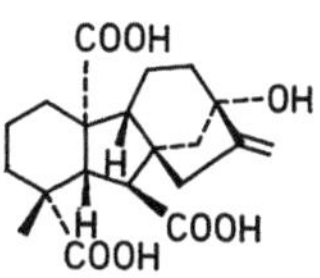	P	$C_{20}H_{26}O_7$ 378	140–150		[32]
GA_{18}	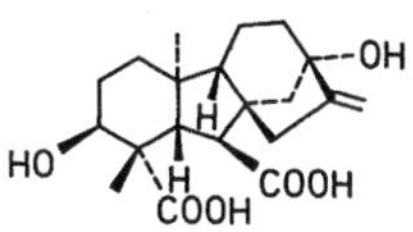	P	$C_{20}H_{28}O_6$ 364	240–242		[33, 34]

Table 1. (Continued)

Compound	Structure	Source F Fungus P Plant	Molecular formular and molecular weight	M. p. in °C	$[\alpha]_D$	References
GA_{19}	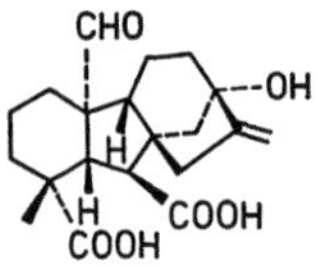	P	$C_{20}H_{26}O_6$ 362	236–237		[35]
GA_{20}	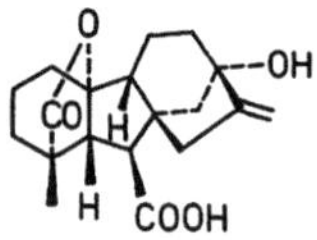	P	$C_{19}H_{24}O_5$ 332	232–233		[36]
GA_{21}	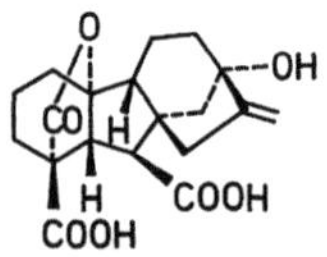	P	$C_{19}H_{22}O_7$ 362	244–246		[37]
GA_{22}	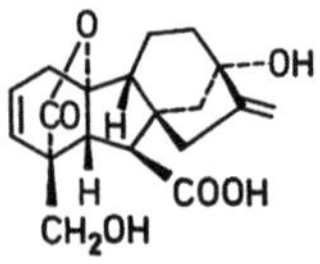	P	$C_{19}H_{22}O_6$ 346			[37]
GA_{23}	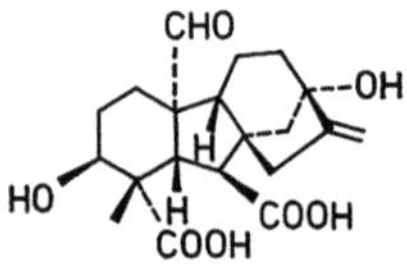	P	$C_{20}H_{26}O_7$ 378	185–189		[38]
GA_{24}	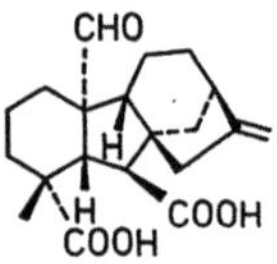	F, P	$C_{20}H_{26}O_5$ 346	199–203	−88°	[39, 40]
GA_{25}	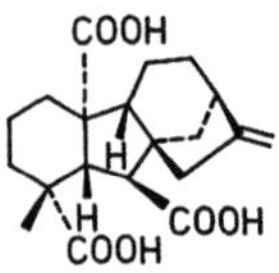	F, P	$C_{20}H_{26}O_6$ 362	154–157 and 248–252	−69°	[39, 40]
GA_{26}	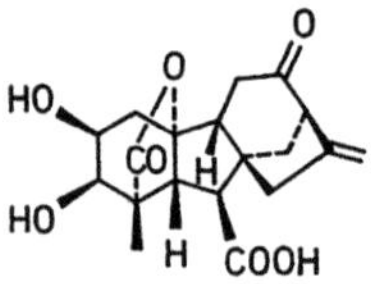	P	$C_{19}H_{22}O_7$ 362	254–257		[41, 42]

Table 1. (Continued)

Compound	Structure	Source F Fungus P Plant	Molecular formula and molecular weight	M. p. in °C	$[\alpha]_D$	References
GA_{27}	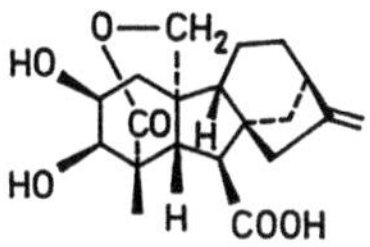	P	$C_{20}H_{26}O_6$ 362	163–165		[41, 42]
GA_{28}	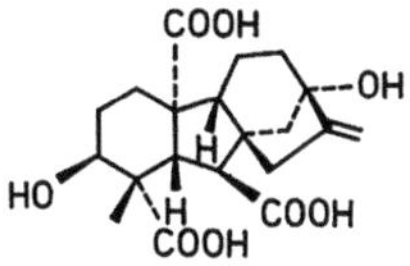	P	$C_{20}H_{26}O_6$ 362	224–225	−7°	[43]
GA_{29}	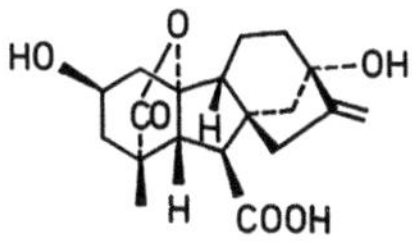	P	$C_{19}H_{24}O_6$ 348	amorphous		[44]
GA_{30}	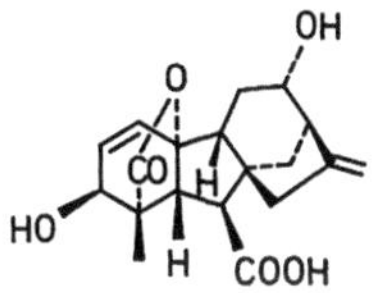	P	$C_{19}H_{22}O_6$ 346	188–191		[45]
GA_{31}	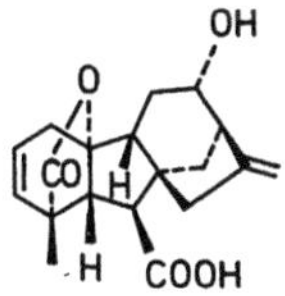	P	$C_{19}H_{22}O_5$ 330	227–231		[45]
GA_{32}	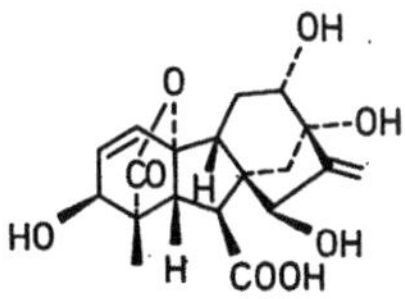	P	$C_{19}H_{22}O_8$ 378			[46]
GA_{33}	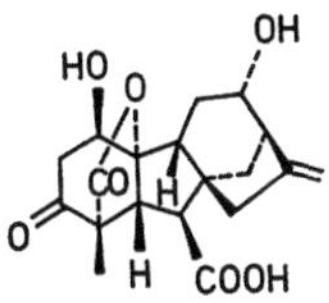	P	$C_{19}H_{22}O_7$ 362	219–221		[47]
GA_{34}	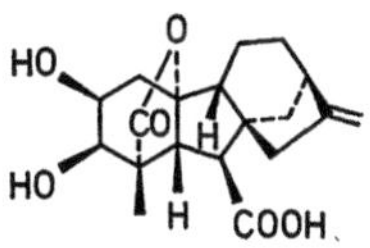	P	$C_{19}H_{24}O_6$ 348	218–219		[47]

Table 1. (Continued)

Compound	Structure	Source F Fungus P Plant	Molecular formula and molecular weight	M. p. in °C	$[\alpha]_D$	References
GA_{35}	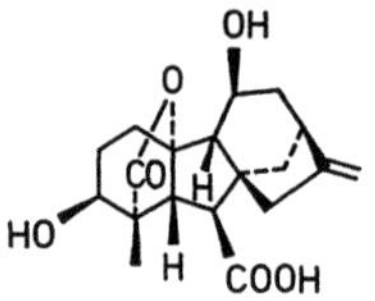	P	$C_{19}H_{24}O_6$ 348	248–252		[48]
GA_{36}	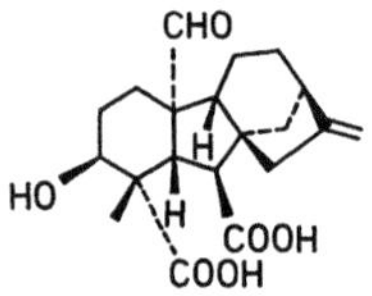	F, P	$C_{20}H_{26}O_6$ 362	205–208		[49, 50]
GA_{37}	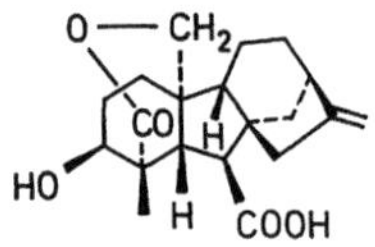	F, P	$C_{20}H_{26}O_5$ 346			[50, 51]
GA_{38}	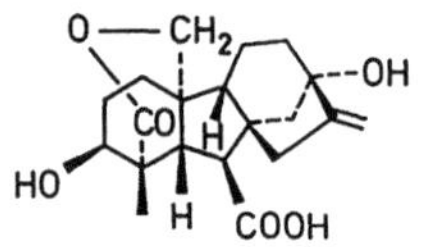	P	$C_{20}H_{26}O_6$ 362			[51]
GA_{39}	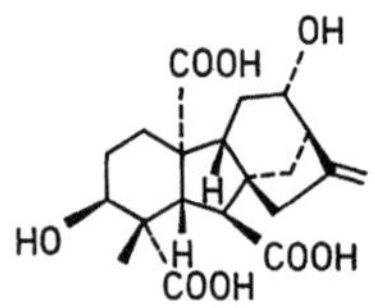	P	$C_{20}H_{26}O_8$ 394			[52]
GA_{40}	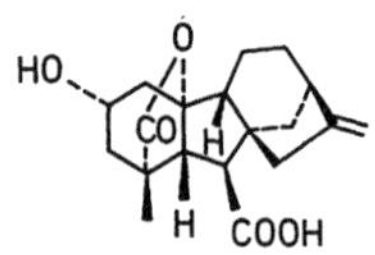	F, P	$C_{19}H_{24}O_5$ 332	212–213		[46, 53]
GA_{41}	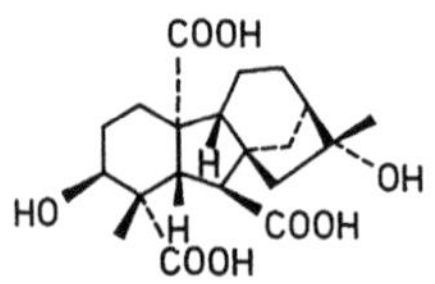	F	$C_{20}H_{28}O_8$ 396			[50]
GA_{42}	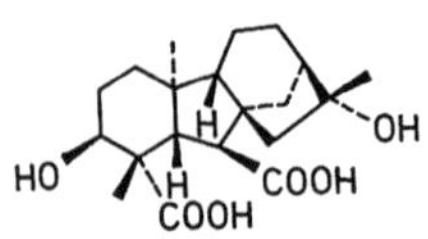	F	$C_{20}H_{30}O_6$ 366			[50]

Table 1. (Continued)

Compound	Structure	Source F Fungus P Plant	Molecular formula and molecular weight	M. p. in °C	$[\alpha]_D$	References
GA_{43}	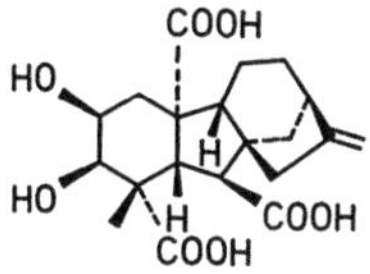	P	$C_{20}H_{26}O_8$ 394			[54]
GA_{44}	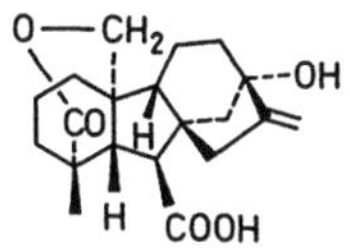	P	$C_{20}H_{26}O_5$ 346			[55]
GA_{45}	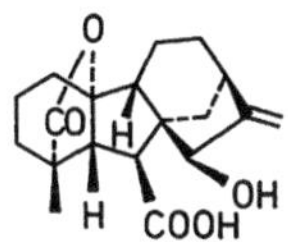	P	$C_{19}H_{24}O_5$ 332			[56]
GA_{46}	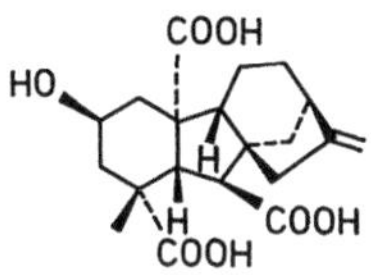	P	$C_{20}H_{26}O_7$ 378			[57]
GA_{47}	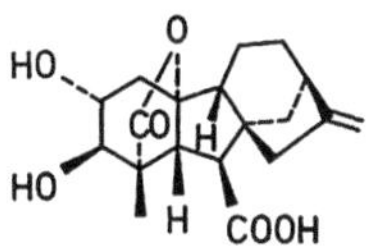	F	$C_{19}H_{24}O_6$ 348			[57]
GA_{48}	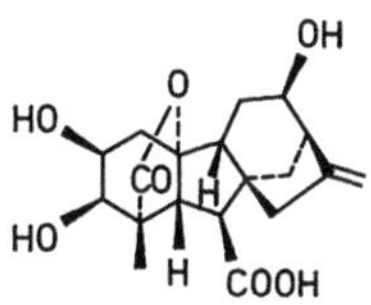	P	$C_{19}H_{24}O_7$ 364		−17°	[52]
GA_{49}	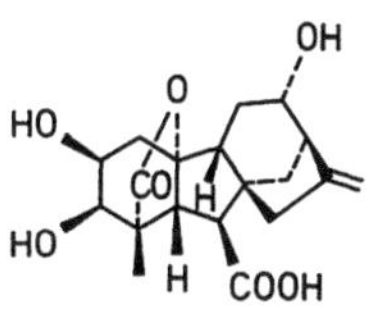	P	$C_{19}H_{24}O_7$ 364		+11°	[52]
GA_{50}	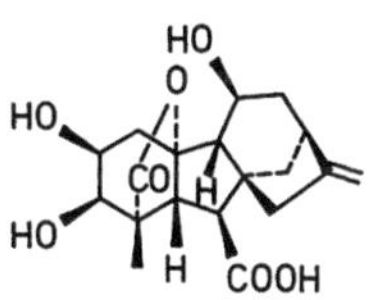	P	$C_{19}H_{24}O_7$ 364			[58]

Table 1. (Continued)

Compound	Structure	Source F Fungus P Plant	Molecular formula and molecular weight	M. p. in °C	$[\alpha]_D$	References
GA_{51}	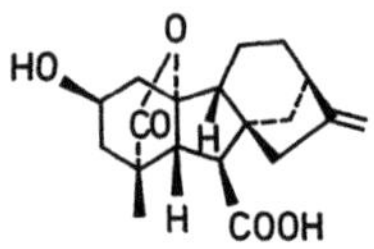	P	$C_{19}H_{24}O_5$ 332			[59]
GA_{52}	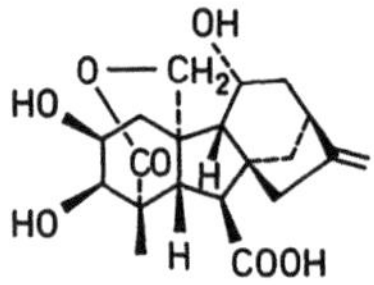	P	$C_{20}H_{26}O_6$ 362			[58]
GA_{53}	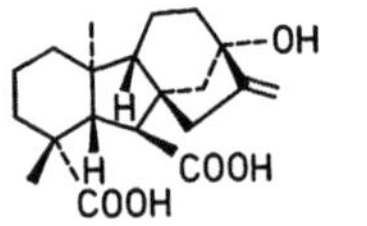	P	$C_{20}H_{28}O_5$ 336			[60]
GA_{54}	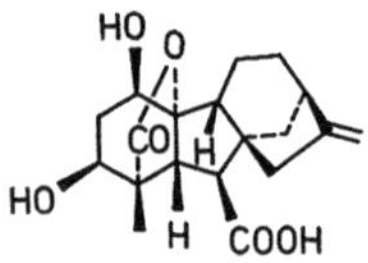	F, P	$C_{19}H_{24}O_6$ 348	243–246		[61, 62]
GA_{55}	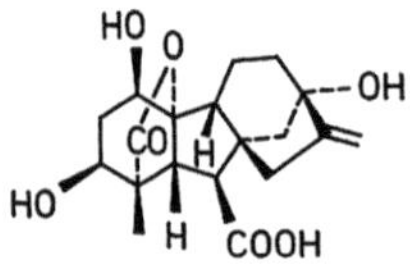	F, P	$C_{19}H_{24}O_7$ 364	260–263	+39°	[61, 62]
GA_{56}	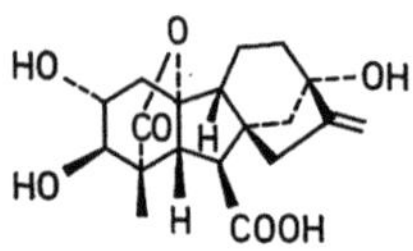	F	$C_{19}H_{24}O_7$ 364			[61]
GA_{57}	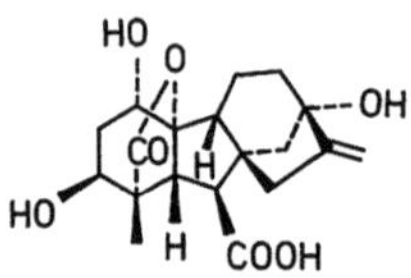	F	$C_{19}H_{24}O_7$ 364	147–150	+19°	[63]
GA_{58}	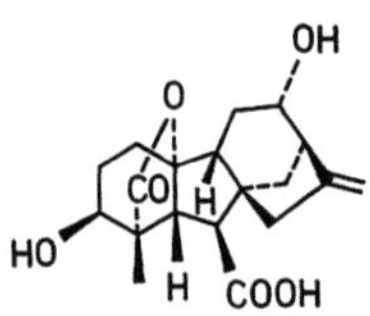	P	$C_{19}H_{24}O_6$ 348			[53, 64]

Table 1. (Continued)

Compound	Structure	Source F Fungus P Plant	Molecular formula and molecular weight	M. p. in °C	$[\alpha]_D$	References
GA_{59}	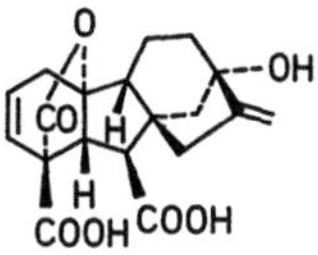	P	$C_{19}H_{20}O_7$ 360			[65]
GA_{60}	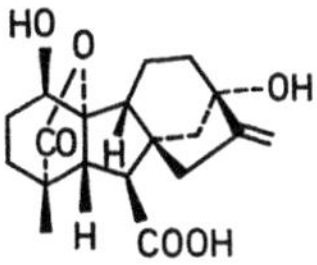	P	$C_{19}H_{24}O_6$ 348			[66]
GA_{61}	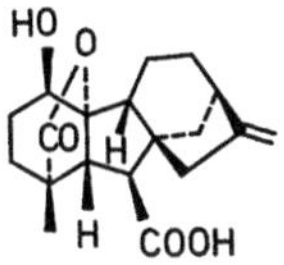	P	$C_{19}H_{24}O_5$ 332	257–259		[66]
GA_{62}	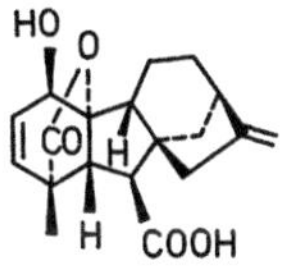	P	$C_{19}H_{22}O_5$ 330			[66]
GA_{63}	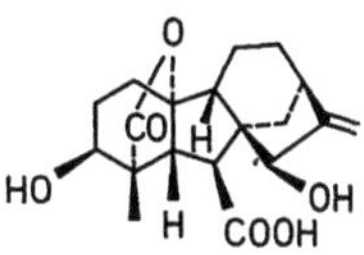	P	$C_{19}H_{24}O_5$ 332			[67]
GA_{64}	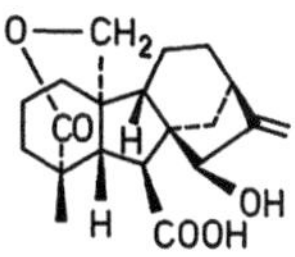	P	$C_{20}H_{26}O_5$ 346	265–267	−67°	[68]
GA_{65}	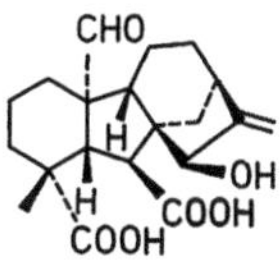	P	$C_{20}H_{26}O_6$ 362		−81°	[68]
GA_{66}	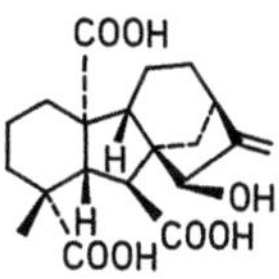	P	$C_{20}H_{26}O_7$ 378			[68]

Table 1. (Continued)

Compound	Structure	Source F Fungus P Plant	Molecular formula and molecular weight	M. p. in °C	$[\alpha]_D$	References
GA_{67}	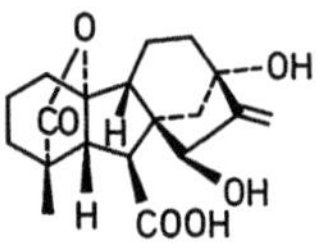	P	$C_{19}H_{24}O_6$ 348	135–137		[69]
GA_{68}	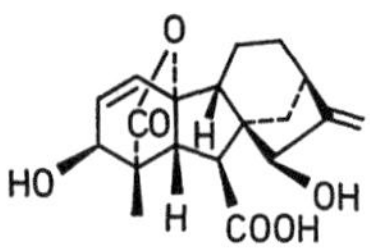	P	$C_{19}H_{22}O_6$ 346			[69]
GA_{69}	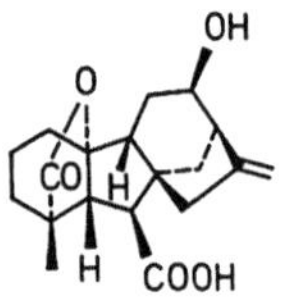	P	$C_{19}H_{24}O_5$ 332			[53, 70]
GA_{70}	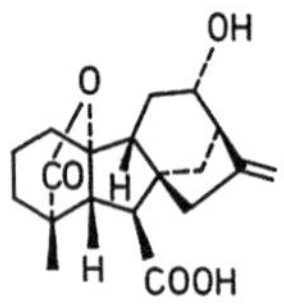	P	$C_{19}H_{24}O_5$ 332			[53, 70]
GA_{71}	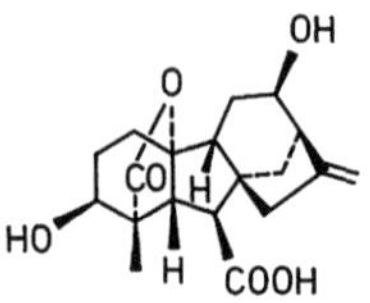	P	$C_{19}H_{24}O_6$ 348			[70]
GA_{72}	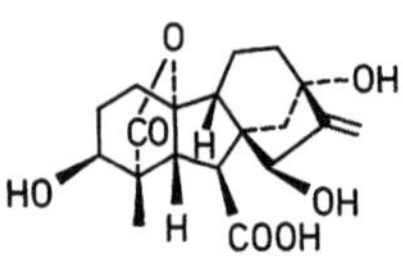	P	$C_{19}H_{24}O_7$ 364			[71]
GA_{73}	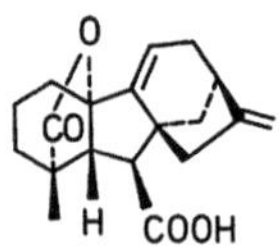	P	$C_{19}H_{22}O_4$ 314			[72]
GA_{74}	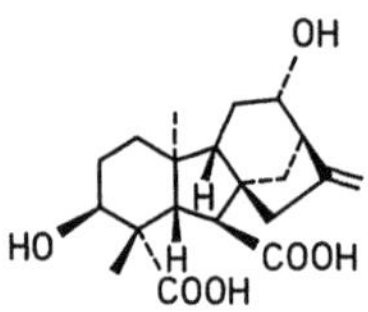	P	$C_{20}H_{28}O_6$ 364			[53, 73]

3 Partial Synthesis of Native Gibberellins

Although remarkable progress has been made in the total synthesis of gibberellins as recently reviewed by Mander [74], the hitherto developed multistep sequences leading to GA_1, GA_3, GA_4, GA_5. GA_{12}, GA_{15} as well as to GA_{38} methyl ester are only of theoretical interest. Therefore, biotechnologically produced GA_3 remains the preferred starting material for the preparation of scarce native gibberellins and structurally modified analogs required for further biological studies [75].

As shown by X-ray diffraction [76], the density of functionality in GA_3 is combined with a considerable strain of the skeleton. Thus, GA_3 represents a highly reactive molecule which undergoes, with the loss of its bioactivity, a series of irreversible reactions, some of them under mild conditions. Even with boiling water GA_3 is transformed to gibberellenic acid (**1**) [77], whereas upon treatment with strong acids aromatization under inversion at C-9 to *allo*gibberic acid (**2**) takes place [78, 79]. More vigorous acid conditions cause simultaneous Wagner-Meerwein rearrangement to give gibberic acid (**3**) [80], a transformation typical of all gibberellins having 13-hydroxy $\Delta^{16(17)}$ olefinic structure. By 0.01 M aqueous NaOH GA_3 undergoes completely lactone rearrangement to *iso*GA_3 (**4**) [81] and with stronger base subsequent lactone opening under formation of the dicarboxylic acid (**5**) [82] occur. Compounds 1, 2 and 4 appear frequently as concomitants of crude GA_3. For all these reasons, specific transformations starting from GA_3 require a careful selection of the reaction conditions.

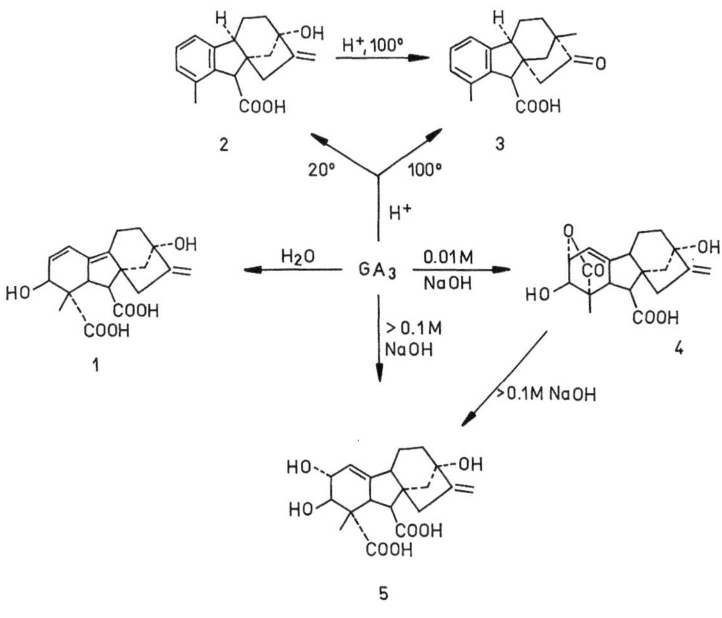

3.1 Gibberellin A_1

GA_1 is available by selective catalytic hydrogenation of GA_3 methyl ester (**6**) in ethyl acetate or THF over a $Pd/CaCO_3$-catalyst in the presence of pyridine [83–86]. However, under these conditions, besides the desired GA_1 methyl ester (**7**) the lactone-opened carboxylic acid **8** was obtained as the main product (ratio 1:2). By relactonization following the Corey procedure [87] the obtained 1β-iodo GA_1 methyl ester (**11**) can be transformed to GA_1 [88, 89] via sodium borohydride reduction to **7** and Bartlett demethylation [88] with lithium *n*-propyl mercaptide (*n*PropLiS). Also GA_3 can be used as starting material for the hydrogenation, although the separation of GA_1 from the simultaneously formed dicarboxylic acid is more difficult in this case. The ring A structure of GA_3 can be reconstructed from the 1β-iodo intermediate **11** by hydrogen iodine elimination with piperidine or photochemically [89, 90].

A disadvantage of the above mentioned hydrogenation step of GA_3 is the over-reduction of the terminale methylene group to form difficult separable tetrahydro compounds. Therefore, several other procedures for GA_1 have been developed. Thus, MacMillan and Willis [91] used the 13-acetoxy derivative of GA_3 for the hydrogenation and reduced the corresponding iodine intermediate with tri-*n*-butylstannane/α,α'-azoisobutyronitril (nBu_3SnH/AIBN). Also the hydrogenation of the 16*S*,17-epoxide **9** [93, 94] obtained from **6** with *m*-chloro perbenzoic acid (*m*ClPBA) via **10** followed by deoxygenation with Zn/AcOH gave **7** [92]. An analogous reaction sequence to GA_1 used the 13-acetoxy derivative of GA_3 for the epoxidation step [93].

Alternatively, the $\Delta^{16(17)}$ double bond of **6** was cleaved by ozonolysis or with osmium tetraoxide/periodate to afford the 16-*nor*ketone **12**, which upon hydrogenation and Wittig reaction with methylene triphenylphosphorane gave GA_1 methyl ester **7** [95]. A non-basic modification of the Wittig step using CH_2Br_2/$TiCl_4$/Zn has been published by Lombardo [96].

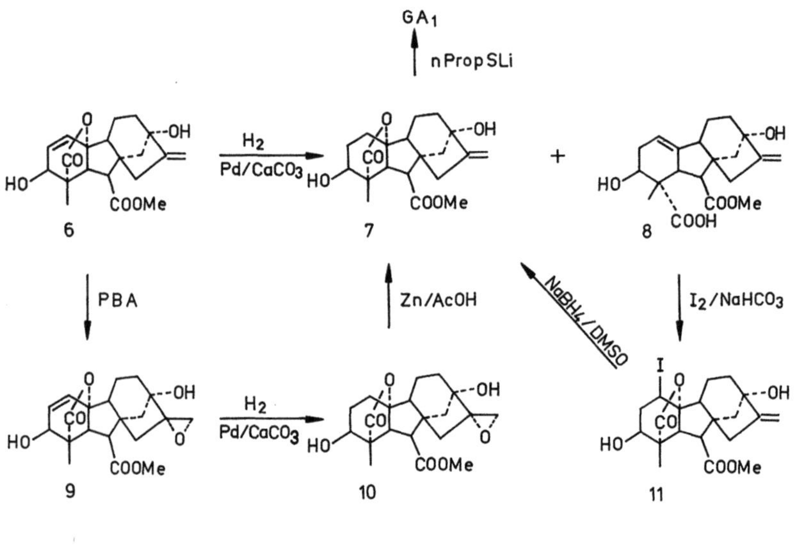

Another key intermediate for the synthesis of GA_1 compounds is 3-dehydro GA_3 (**13**) and its esters respectively, available via MnO_2 oxidation of GA_3 under Attenburrow conditions [97, 99] or with pyridinium dichromate (PDC) in DMF [100]. Whereas the borohydride reduction of such enones under various modifications gave predominantely the unnatural 3α-epimers e.g. **14** [100, 101], the reaction of **13** with K-selectride ($\mathit{Ki}Am_3BH$) led to the desired GA_1 as the main product [102].

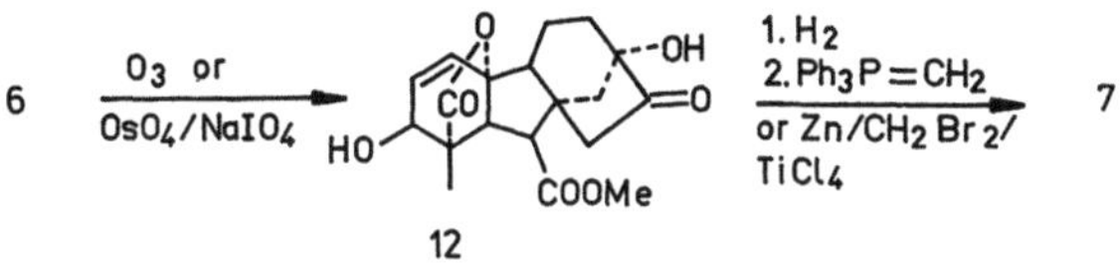

12

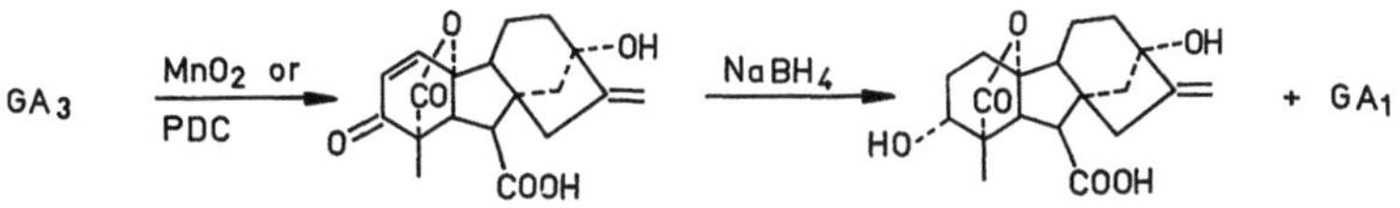

13 14

3.2 Gibberellin A_2

GA_2 was obtained by acid catalyzed (3N HCl) hydration of GA_4 [103]. This reaction proceeds stereospecifically under formation of the native 16(*R*)-stereochemistry based on the water addition from the less hindered face of the molecule [104]. On the other hand the terminal methylene function can be reconstituted by dehydration as shown in the reaction of 3-acetoxy-GA_2 methyl ester with $POCl_3$/pyridine to give the corresponding GA_4 derivative [14].

3.3 Gibberellin A_4

GA_4 is available from GA_1 by selective deoxygenation of the 13-hydroxy group. Thus, the GA_1 derivative **15** was reacted with methyl oxalyl chloride to give the ester **16** which upon heating with tri-*n*-butyl stannane/α,α′-azoisobutyronitrile in benzene afforded the GA_4 ester **18** [105]. This ester was also obtained by reaction of **15** with Ph_3P/CCl_4/pyridine to the 13-chloro compound **17** followed by stannane reduction [106]. Bartlett demethylation and hydrolysis led to GA_4.

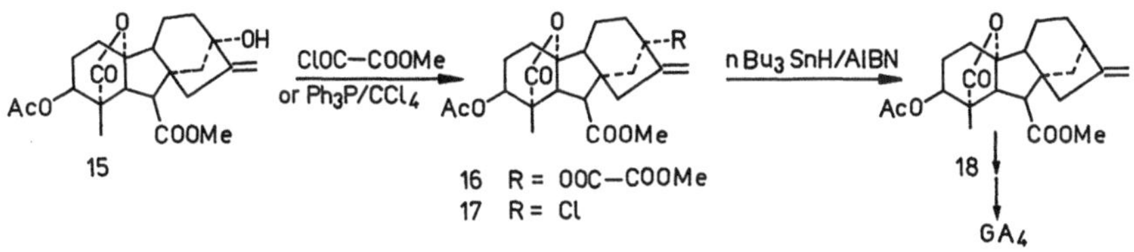

In analogy to the transformation of GA_3 into GA_1 via the lactone opened intermediate **8** GA_4 is also obtainable by hydrogenation of GA_7 and relactonization [86].

A conversion of GA_{13} into GA_4 was simultaneously published by two groups [86, 107] and involves the oxidative decarboxylation of C-20 carboxyl with lead tetraacetate followed by relactonization as key reactions.

3.4 Gibberellin A_5

Several procedures for the preparation of GA_5 have been published starting either from suitable GA_1 or GA_3 derivatives. Thus, GA_5 methyl ester (**24**) was prepared by elimination of the 3*β*-monotoluene-*p*-sulphonate **19** with collidine [13, 20]. Although the elimination reaction of corresponding equatorial 3*α*-derivatives is more difficult, the yields can be improved by addition of nBu_4NBr [108]. Similar, the dehydration of the GA_1 diester **21** with $POCl_3$/pyridine gave the 13-acetate of GA_5 methyl ester. Hydrolysis with 2N NaOH/MeOH under reflux led directly to GA_5 [109]. Also GA_1 methyl ester (**7**) has been used in the reaction with Ph_3P/CCl_4/pyridine, although in that case besides the desired GA_5 methyl ester (**24**) the 3,13-dichloro compound of **7** was formed as the main product [100].

As a further route Murofushi et al. [92] found, that the selective hydrogenation of GA_3 methyl ester monomesylate (**20**) with $Pd/BaCO_3$ in ethyl acetate/pyridine afforded under migration of the Δ^1 double bond also GA_5 methyl ester (**24**).

All the mentioned pathways for GA_5 have disadvantages in regard to poor yields and formation of byproducts. A more improved reaction sequence with an overall yield of 20% GA_5 from GA_3 was developed by Bearder et al. [110]. In this

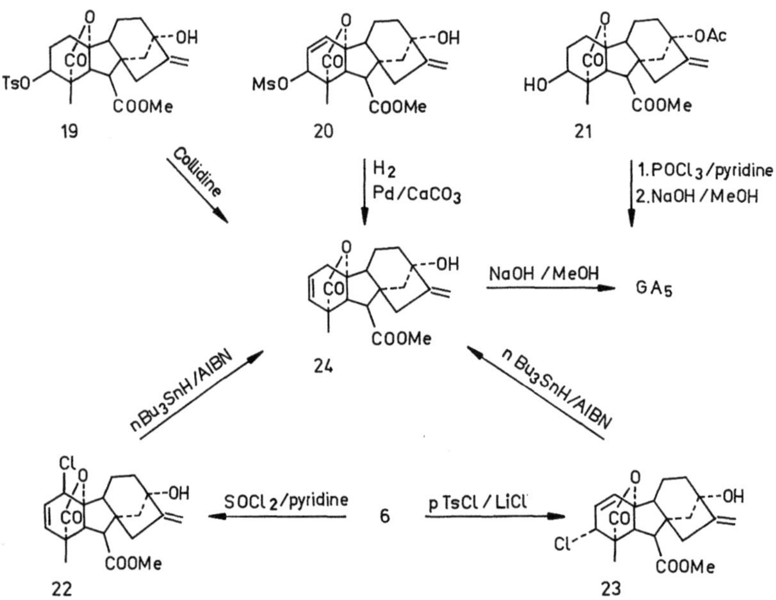

procedure GA_3 methyl ester (**6**) was reacted with $SOCl_2$/pyridine to give the 1β-chloro GA_5 methyl ester (**22**). On the other hand methyl ester **6** with *p*TsCl/LiCl led to the isomeric Δ^1-3α-chloro compound **23**. Both chloro compounds were reduced by *n*Bu_3SnH/AIBN to the GA_5 methyl ester (**24**). The chloro compounds **22** and **23** can be also prepared by treatment of **6** with Ph_3P/CCl_4 [111] or from the mesylate **20** with LiCl [112].

3.5 Gibberellin A_7

In analogy to the partial synthesis of GA_4 from GA_1 also GA_7 was obtained from GA_3 by deoxygenation of the 13-hydroxy function. Thus, in a patent [113] a procedure for the preparation of GA_7 has been described which involves the transformation of 3-acetoxy-GA_3 **25** with MsCl/pyridine to the mesylated acid chloride **28**, followed by treatment with $NaHCO_3$ to give the acid **29**. Subsequent reduction with *n*Bu_3SnH/AIBN to **32** and hydrolysis with K_2CO_3 led to GA_7. Reaction of diester **26** with methyl oxalyl chloride and stannane reduction of the so obtained **30** gave 3-acetoxy-GA_7 methyl ester **33**. A similar reaction sequence was achieved starting with the 3-acetate of GA_3 [105].

Recently, an effective synthesis of GA_7 was elaborated using the phenacyl ester **27** as a key intermediate, which was transformed with Ph_3P/CCl_4/pyridine to the 13-chloro compound **31**. The stannane reaction of **31** allowed simultaneously a reductive elimination of the chlorine as well as the cleavage of the phenacyl ester function leading to 3-acetoxy GA_7 **32** [106].

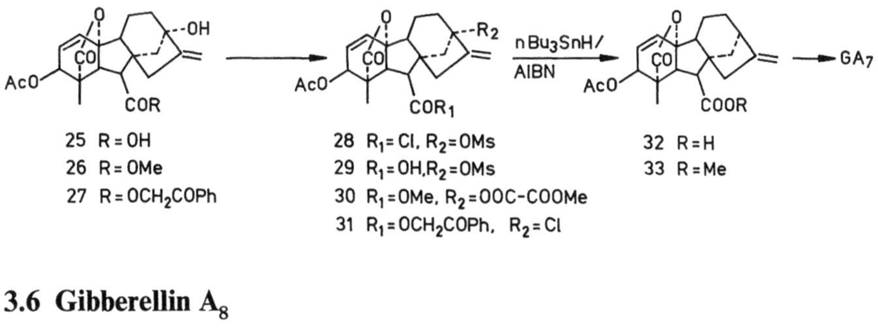

25 R = OH
26 R = OMe
27 R = OCH_2COPh

28 R_1 = Cl, R_2 = OMs
29 R_1 = OH, R_2 = OMs
30 R_1 = OMe, R_2 = OOC-COOMe
31 R_1 = OCH_2COPh, R_2 = Cl

32 R = H
33 R = Me

3.6 Gibberellin A_8

GA_8 as a member with 2β,3β-glycol function was prepared by hydroxylation of GA_5 with OsO_4/dioxane besides other products, probably 13,16,17-trihydroxy- and 2,3,13,16,17-pentahydroxy-derivatives [92].

3.7 Gibberellin A_9

The preparation of GA_9 requires the removal of both 3,13-hydroxy functions from suitable GA_1 derivatives. Thus, GA_9 was obtained from 3-*epi* GA_1 methyl ester (**34**) (see 3.1.) by treatment with Ph_3P/CCl_4/pyridine to give the 3β,13-dichloro compound **35**, which upon stannane reduction afforded the GA_9 methyl ester and further, via alkaline hydrolysis, GA_9. In a similar way 3-*epi* GA_4 methyl ester (**36**) was converted with $POCl_3$/pyridine to the 3β-chloro compound **37** and further transformed to GA_9 [109]. Also a mixture of GA_4/GA_7 has been used for the preparation of GA_9 [100]. Starting from GA_4 methyl ester (**38**) the deoxygenation of the 3β-hydroxy group was achieved via the methyl oxalyl ester **39** [105].

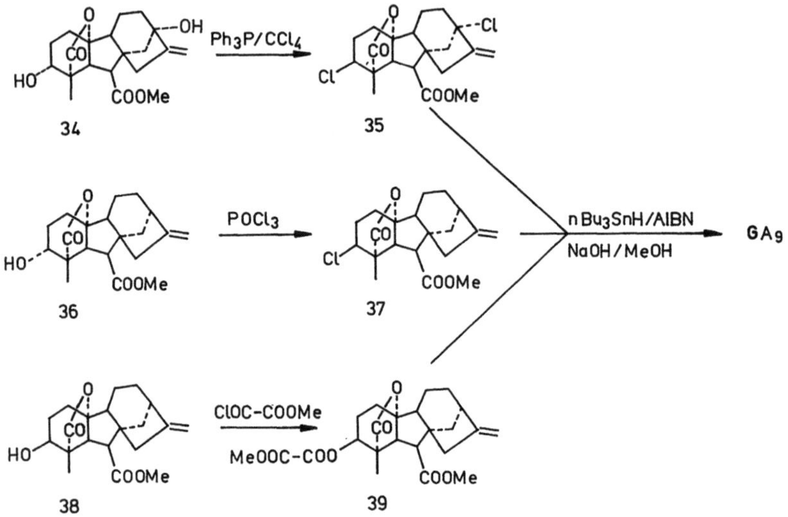

3.8 Gibberellin A_{10}

GA_{10} is available from GA_9 by acid catalyzed hydration of the terminal methylene function similar as described for the preparation of GA_2 from GA_4 (see Sect. 3.2) [14].

3.9 Gibberellin A_{19}

The synthesis of GA_{19} from GA_3 was published by Dawe et al. [114, 115] representing the first conversion of a C_{19}-into a C_{20}-gibberellin. The 3,13-bismethoxymethyl ether **40** was reduced with Li/liq. NH_3 to the 3-deoxy carboxylic acid **41**, which was transformed with oxalyl chloride/pyridine to the acid chloride **42** and further reacted with CH_2N_2 to give the corresponding diazoketone **43**. Cyclopropanation of **43** with copper-bronze led to the cyclopropylketone **44**, the C(19)-C(20)-bond of which was reductively opened with Li/liq. NH_3 to give **45**. Treatment with KH/DMF/MeI furnished the enol ether **46**, which afforded upon selective oxidative cleavage with ozone at −78 °C and treatment with Me_2S the GA_{19} derivative **47**. Removal of the 13-methoxymethyl function with traces of Me_3SiCl in $CHCl_3$/MeOH and final hydrolysis by refluxing with 20% KOH gave GA_{19}.

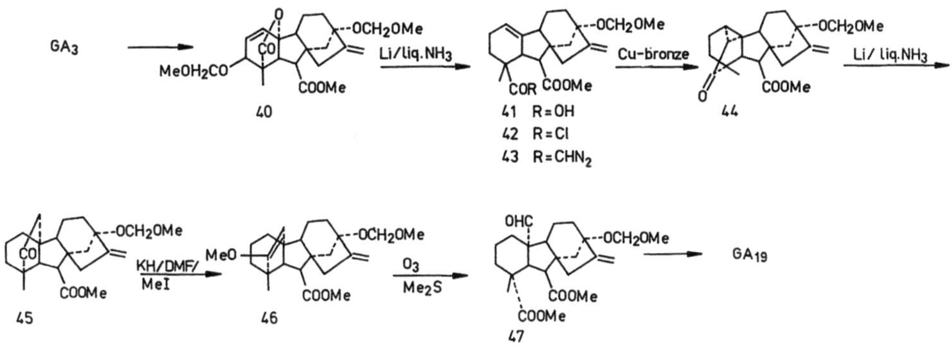

3.10 Gibberellin A_{20}

GA_{20} was first obtained in low yield by selective catalytic hydrogenation of GA_5 methyl ester (**24**) and alkaline hydrolysis [36]. Later an improved procedure has been published by the same group which involved protection of the terminal methylene function of **24** to the epoxide **48** followed by catalytic hydrogenation to give **49**, regeneration of the $\Delta^{16(17)}$ double bond and hydrolysis [92]. In other procedures the stannane reduction of the 3β-chloro derivative **52** [100] or the 3α-thiobenzoate **51** [109], both available from 3-*epi*-GA_1 methyl ester acetate **50**, was used. Also the stannane reduction of the corresponding methoxalyl ester **53** has been published [105] as a method for GA_{20} synthesis.

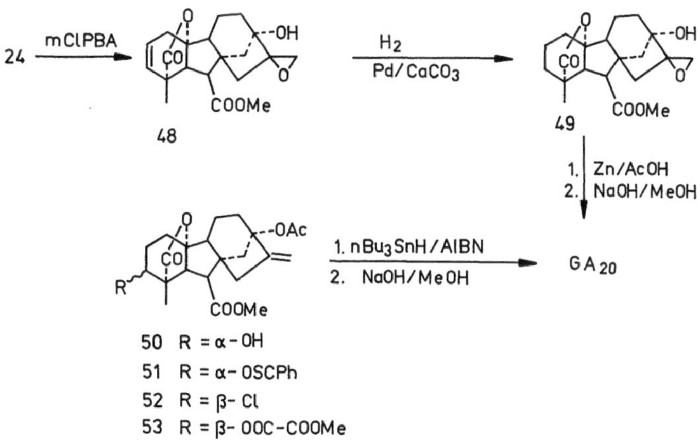

3.11 Gibberellin A_{25}

This C_{20} gibberellin is available from the fungal GA_{13} in a reaction sequence as follows: Starting with the trimethylester **54** the Δ^2-*nor*ketone **55** was prepared via oxidative cleavage of the terminal methylene function and introduction of the Δ^2 double bond by refluxing of the corresponding 3-tosylate with collidine.

Hydrogenation over PtO_2 in AcOH and Wittig reaction with methylene triphenyl phosphorane led to GA_{25} trimethyl ester **56** [40].

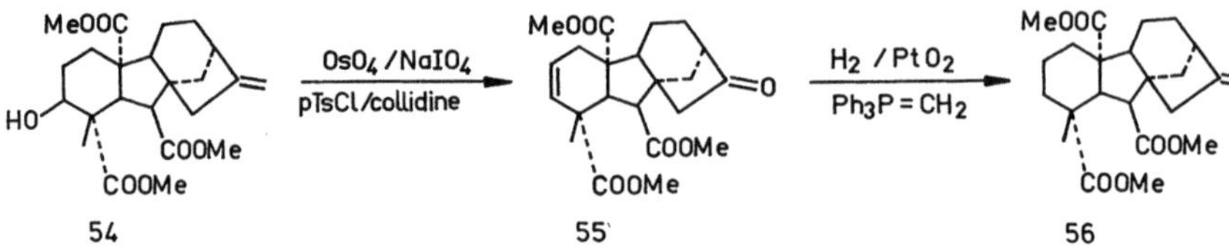

3.12 Gibberellin A_{29}

Several pathways leading to labelled GA_{29} have been published and are described in Sects. 4.1.3, 4.2.2 and 4.2.3.

3.13 Gibberellin A_{31}

An original reaction sequence to 12-hydroxylated gibberellins has been developed recently by Chu and Mander [116]. Applied to the synthesis of GA_{31} the 13-acetate of GA_5 methyl ester (**57**) was hydroxylated at C-15 and further oxidized with PCC to give both aldehydes **58** and **59** in a 2:1 ratio. The mixture was reduced with $CrCl_2$ to the aldehyde **60**, which gave upon bromination, followed by $NaBH_4$-reduction the bromohydrine **61**. The desired 12-function was introduced by reaction with $Pb(OAc)_4$/iodine under irradiation leading to the ether **62**, which underwent with Zn/AcOH ring opening and HBr elimination to the 12β-carbinol **63**. To obtain the native 12α-configuration, **63** was oxidized and hydrolysed to

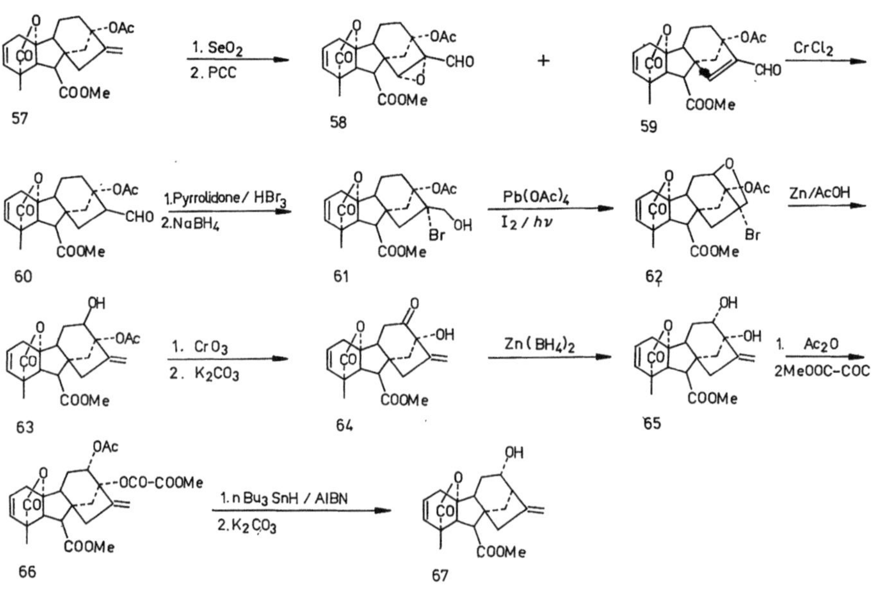

the ketol **64**, which exhibited, because of chelation, a conformation suitable for a preferred nucleophilic attack to the 12-oxo function from the upper face. Thus, reduction of **64** with Zn $(BH_4)_2$ in DMF led to a 3:2 mixture of the desired 12α-compound **65** and its 12β-epimer. Selective acetylation, transformation of the obtained 12α-acetate to the 13-methyl oxalate **66** and stannane reduction gave GA_{31} methyl ester **67**. Preliminary experiments showed, that this strategy can be used also for the synthesis of GA_{30} and GA_{32}.

3.14 Gibberellin A_{34}

A general method of the conversion of 3-hydroxy gibberellins into the methyl esters of 2-hydroxy gibberellins was described by Beely and MacMillan [57]. For the synthesis of GA_{34} methyl ester (**71**) the norketone **68**, prepared by oxidation of a 7:3 mixture of GA_4 and GA_7 with $OsO_4/NaIO_4$ and methylation, was dehydrated to the Δ^2-olefin **69** and hydroxylated with OsO_4 to the diol **70**. Regeneration of the terminal methylene group with methylene triphenylphosphorane gave **71**.

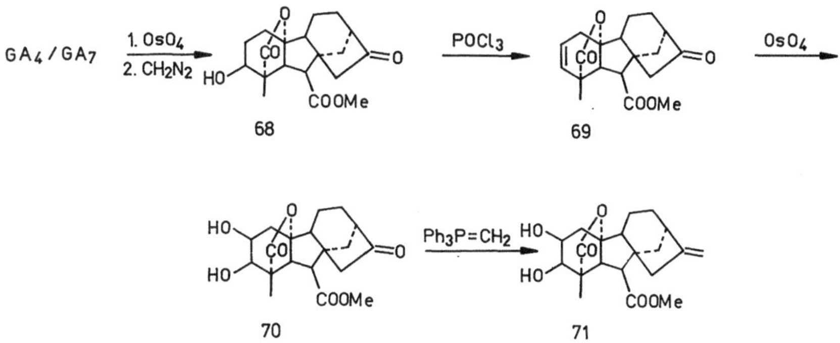

3.15 Gibberellin A_{36} and A_{37}

The synthesis of the GA_{36} and GA_{37} methyl esters from GA_3 follow a similar strategy as described for GA_{19} [115] (see Sect. 3.9). Thus, the 3,13-diacetate **72**, obtained from compound **8** was transformed to the diazoketone **73** followed by intramolecular cyclopropanation to the ketone **74**. Treatment with Li/liq. NH_3 gave mainly the 13-deoxy compound **75**. As an improved procedure for the oxidative fission of C(19)-C(20)-bond **75** was transformed to the *t*-butyldimethylsilyloxy derivative **76** which was enolized with KH, oxidatively cleaved and methylated to give the aldehyde **79**. Removal of the protecting silyl group with nBu_4NF afforded the desired GA_{36} dimethyl ester **80**. Alternatively, the methoxymethyl derivative **77** was used for the oxidative cleavage to give the aldehyde **81** which led upon treatment with dimethyl boron bromide at —78 °C to the desired **80** with improved yield.

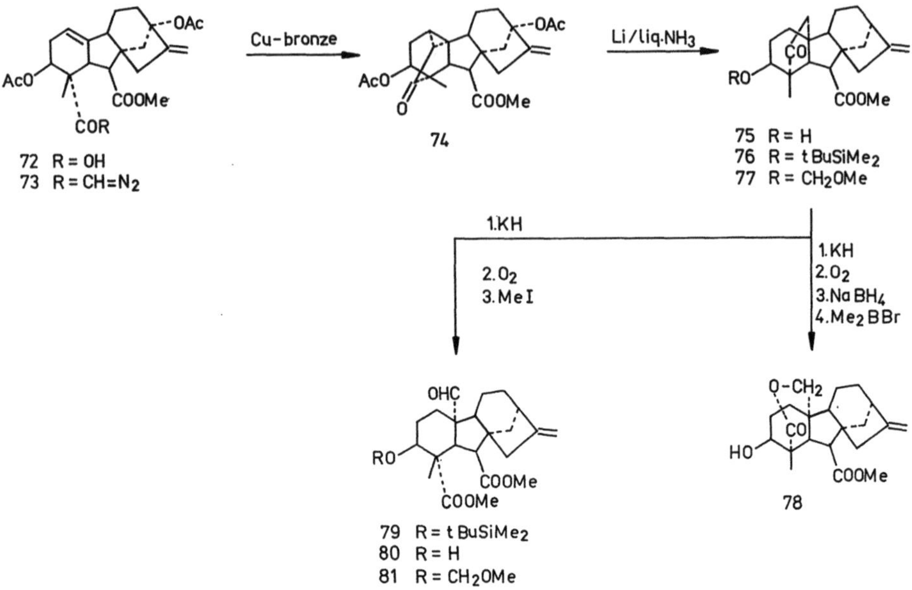

For the preparation of GA_{37} methyl ester (**78**) the aldehydo carboxylate, obtained from **77** by oxidative cleavage without subsequent methylation, was reductively lactonized with $NaBH_4$ and deprotected.

3.16 Gibberellin A_{40}

For the synthesis of the 2α-hydroxylated member GA_{40} the Δ^2-*nor*-ketone **69** (see Sect. 3.14) was reacted with *N*-bromoacetamide to give regio-and stereoselectively the bromo acetate **82**. Debromination and reconstruction of the terminal methylene function followed by alkaline hydrolysis led to GA_{40} methyl ester (**83**) [57].

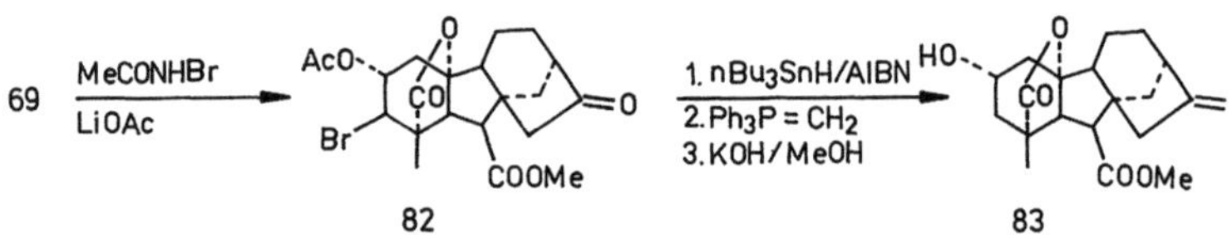

3.17 Gibberellin A_{45}

Dolan et al. [67] described the synthesis of the 15β-hydroxylated GA_{45} from GAg by oxidation of its phenacyl ester (**84**) with SeO_2/*t*BuOOH to the 15α-substituted derivative **85**, followed by Swern oxidation with oxalyl chloride/DMSO/*i*$Prop_2$EtN to give enone **86**. Subsequent reduction with Zn/AcOH under simultaneously deprotection furnished GA_{45}.

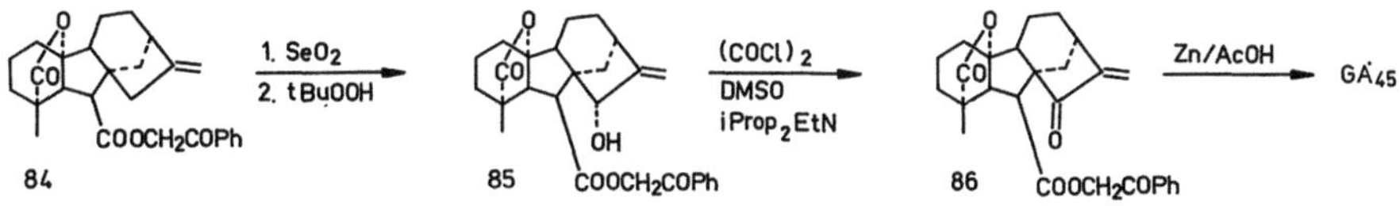

3.18 Gibberellin A_{46}

For the synthesis of GA_{46} trimethyl ester (**91**) from GA_{13} the intermediate **55**, also used for the preparation of GA_{25} (see Sect. 3.11), was reacted with *N*-bromo-acetamide to the bromo hydrine **87**. Protection of the 2α-hydroxy function by silylation with Me_3SiCl to **88** and stannane reduction furnished the debromo compound **89**. Upon reconstruction of the terminal methylene function by Wittig reaction followed by Jones oxidation the 2-ketone **90** was prepared. In the following reduction step optimal results were obtained with in-situ prepared aluminium *iso*propoxide leading to the desired GA_{46} methyl ester (**91**) together with only 10% of the 2α-epimer [57].

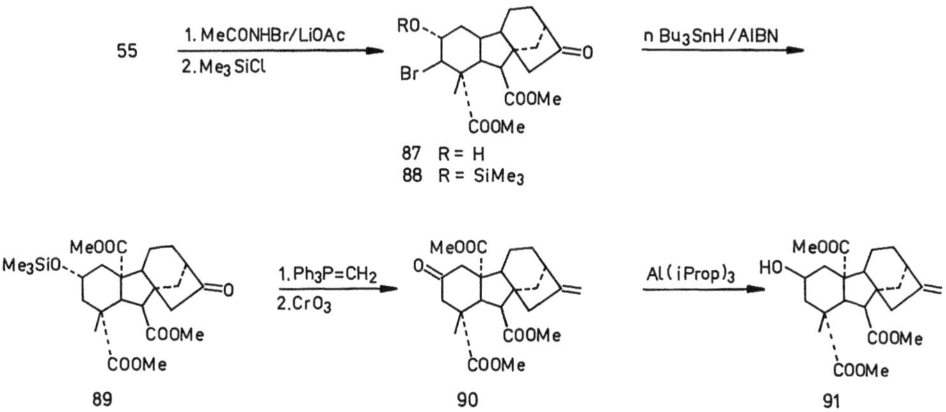

3.19 Gibberellin A_{47}

For the synthesis of GA_{47} the bromoacetate **82** used also as intermediate in the preparation of GA_{40} (see Sect. 3.16) was transformed to the 2α,3α-epoxide **92** which afforded upon treatment with H_2SO_4/THF a mixture of epimeric 2,3-diols. The bis-trimethyl silyl derivative of the separated 2α,3β-diol **93** was used for the Wittig reaction with methylene triphenylphosphorane leading after subsequent removal of the silyl groups to GA_{47} methyl ester (**94**) [57].

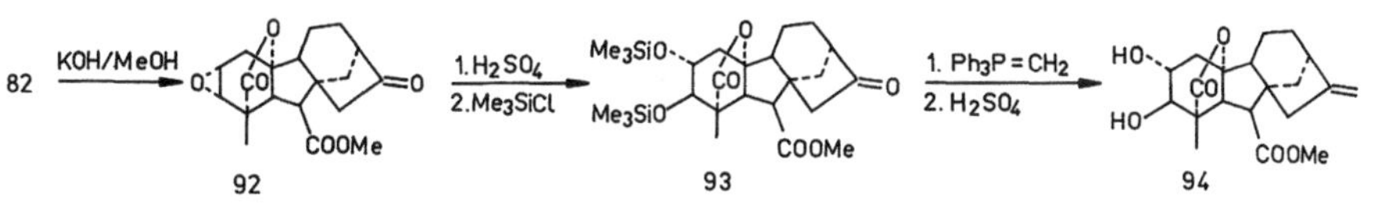

3.20 Gibberellin A_{51}

A reaction sequence for preparing labelled GA_{51} from GA_3 via GA_{40} methyl ester was published as described in Sect. 4.2.3. Using a new approach for the synthesis of 2β- and 3β-hydroxy gibberellins via S_N2 displacement of 2α- and 3α-methanesulphonates with caesium acetate Willis [117] transformed the GA_{40} derivative **95** quantitatively to the 2β-acetate **96**. Alkaline hydrolysis with 2N NaOH under reflux gave GA_{51}. In lower yield the transformation of the mesylate **95** to GA_{51} was achieved also with KO_2/18-crown-6.

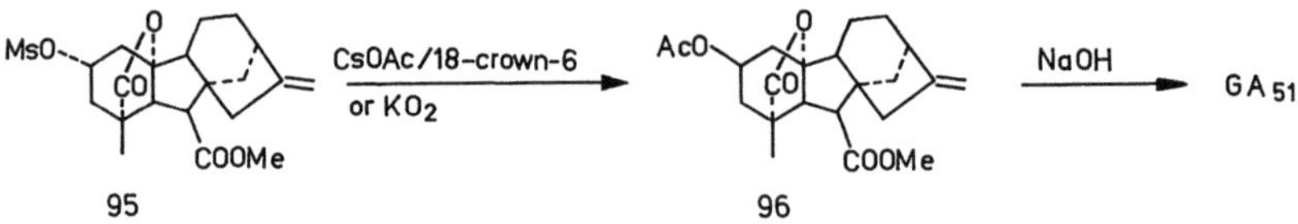

3.21 Gibberellin A_{54}

GA_{54} was obtained in low yield by partial hydrogenation of GA_7 to give the $\Delta^{1(10)}$ carboxylic acid **97** which was treated with *m*ClPBA in dioxane [61].

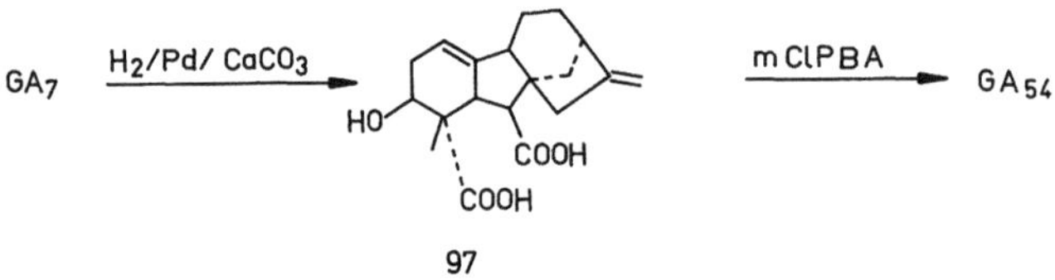

3.22 Gibberellins A_{55} and A_{57}

Using the same principle as described for GA_{54} (see Sect. 3.21) GA_{55} was prepared from GA_3 by reaction of its hydrogenolysis product with *m*ClPBA [61]. Also the corresponding monomethyl ester **8** (see, 3.1) was used for this reaction sequence.

A more improved procedure leading to 1-hydroxylated gibberellins has been published by Adam and Hung [118] and involved the direct hydration of the enone acid **13** to give smoothly the 1β-hydroxy keto acid **99** besides small amounts of its 1α-epimer **98**. $NaBH_4$ reduction of **99** afforded GA_{55} and the corresponding 3α-compound in a 3:1 ratio. Later the same reaction sequence was used for the preparation of GA_{55} and GA_{57} methyl esters starting from the corresponding methylate of **13** [63].

Voigt and Adam described an alternative synthesis of GA_{55} and GA_{57} from **13**, in which the nucleophilic addition of hydrazoic acid to the Δ^1 double bond was applied as key reaction for 1-functionalization. The so obtained 1:3 mixture of epimeric azido ketones **100** and **101** upon $NaBH_4$ reduction gave the corresponding azido alcohols **102**–**105** in a 4:2,5:2,5:1 ratio. The transformation of the azido group to the 1-oxo function was achieved by photolysis (254 nm) of the azido alcohols to give due to loss of nitrogen the corresponding instable imines which were smoothly hydrolyzed to the 1-oxo-3-hydroxy gibberellins **106** and **107**. Subsequent reduction of the mixture and SiO_2 chromatography led to GA_{55} and GA_{57} as well as their 3α-stereoisomers.

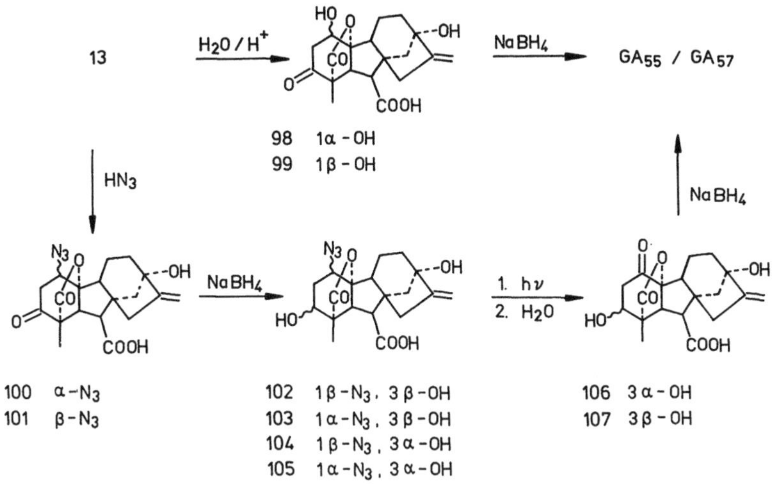

3.23 Gibberellin A_{56}

GA_{56} was prepared by alkaline treatment (pH 13) of GA_3 to give the dicarboxylic acid **5**, followed by relactonization and stannane reduction of the intermediate 1-iodo compound **108** [61].

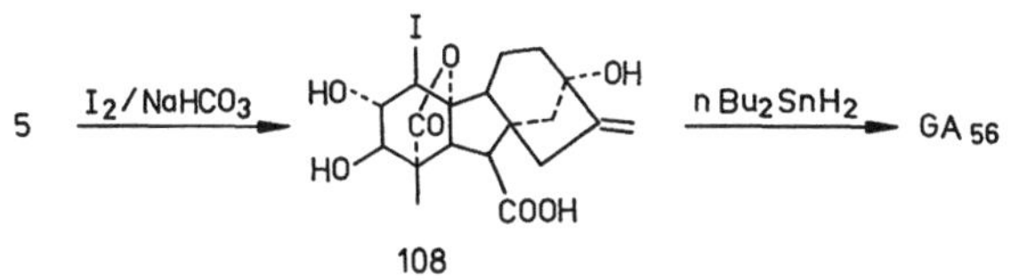

3.24 Gibberellins A_{60} and A_{61}

For the preparation of GA_{60} and GA_{61} in a procedure published by Kirkwood et al. [66] the 1β-hydroxy ketone **109**, obtained by hydration of the methyl ester of enone acid **13**, was acetylated with Ac_2O/*p*TsOH to give the 1β,13-diacetate which afforded upon reduction with $Na(MeO)_3BH$ the 3-epimeric alcohols **110** and **111** in a 1:2 ratio. Thiobenzoylation of this mixture led to the 3α-thiobenzoate **112** besides unreacted 3β-alcohol **111**, the latter one can be hydrolyzed to GA_{55}. Reduction of **112** with *n*Bu_3SnH, prepared from (tri-*n*-butylstannyl)oxide and polymethyl hydrosiloxane, afforded the diacetate **113**, which was hydrolyzed in two steps to give GA_{60}. In an analogous reaction sequence the 13-desoxy gibberellin GA_{61} was prepared from GA_7.

In an alternative pathway the dehydration of both 3-epimeric 1-ketones **106** and **107** with Ac_2O/pyridine afforded smoothly the enone **114**, which was hydrogenated with 10% $Pd/CaCO_3$ to the 1-oxo-gibberellin **115**. $NaBH_4$ reduction led to GA_{60} and its 1α-epimer in a 3:5 ratio [119].

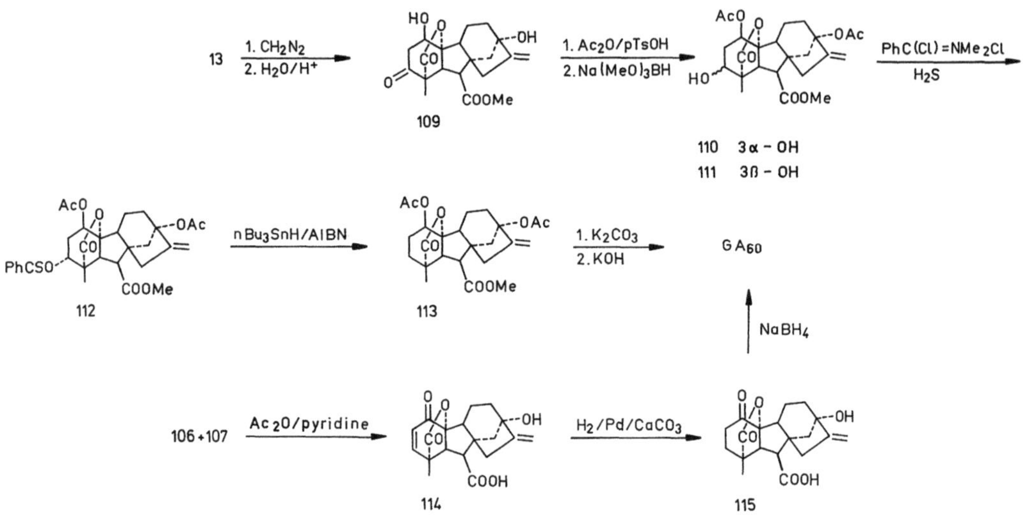

3.25 Gibberellin A_{62}

GA_{62} was obtained from GA_7 via the enone ester **116**, which afforded upon acid catalyzed hydration, diacetylation and reduction with $Na(MeO)_3BH$ a mixture of 3-epimeric alcohols **117** and **118**. Subsequent treatment of this mixture with $POBr_3$/pyridine and 1,5-diaza-bicyclo-/5,4,0/-undec-5-ene (DBU) gave a 17-bromo-15-ene as the main product besides the desired Δ^2-olefin **119**. Alkaline hydrolysis of the mixture gave GA_{62} in low yield [66].

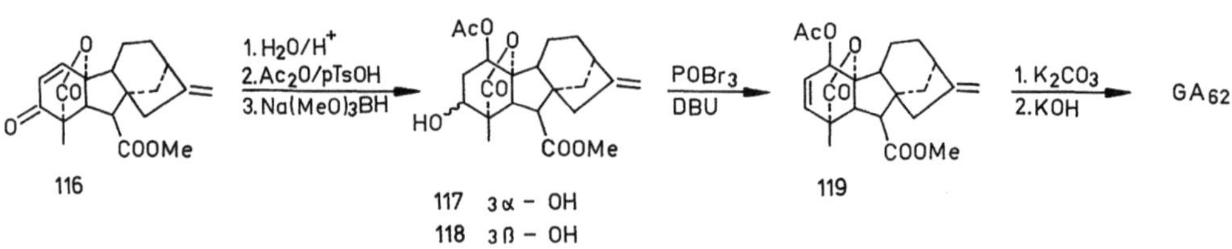

3.26 Gibberellins A_{63}, A_{67} and A_{68}

Similarly to the described synthesis of GA_{45}, the 15β-functionalized GA_{63} was prepared from GA_4 by reaction of the phenacyl ester 3-acetate with SeO_2/*t*BuOOH to the 15α-hydroxy compound **120**, which upon Swern oxidation to the 15-oxo derivative followed by reduction with Zn/AcOH and hydrolysis afforded GA_{63}

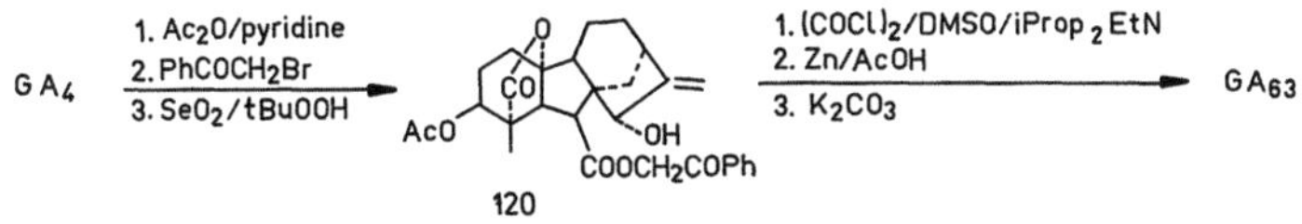

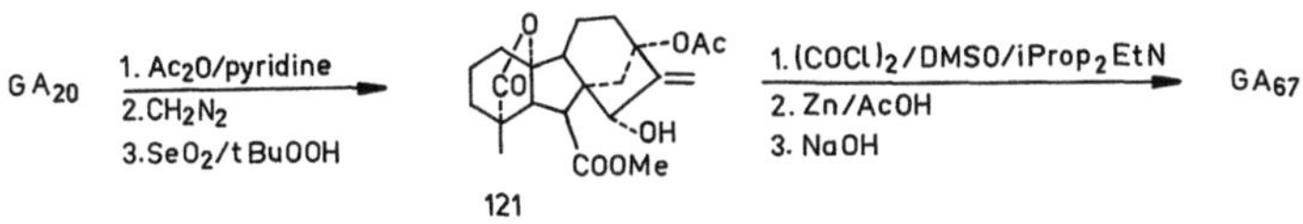

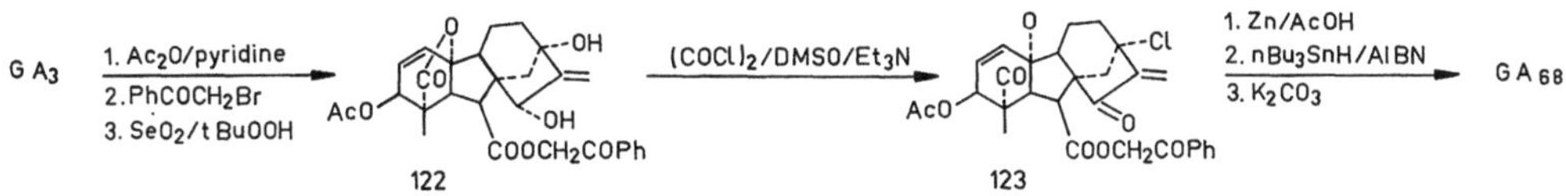

[67]. An analogous reaction sequence starting from GA_{20} led via a 15α-hydroxy methyl ester **121** to GA_{67} [69].

GA_{68} was synthesized from GA_3 using a modified Swern oxidation procedure of the corresponding 15α-hydroxy ester **122**. Thus, upon reaction with $(COCl)_2$/DMSO in presence of triethyl amine instead of *i*$Prop_2EtN$ simultaneous nucleophilic displacement at C-13 led to the 13-chloro ketone **123**, which was reduced in two steps and hydrolyzed to give GA_{68} [69].

3.27 Gibberellins A_{69} and A_{70}

In a preliminary communication Chu and Mander [116] described the preparation of the methyl esters of GA_{69} and GA_{70} in an analogous manner as shown for GA_{31} involving the ring A saturated bromo ether **62** (see Sect. 3.13) as the key intermediate.

4 Labelling of Gibberellins

Labelled gibberellins have been used extensively to investigate transport, localization, metabolism and mode of action of this phytohormone group in plants [1]. In additional, such compounds are applied for investigation of the biosynthesis and for RIA tracers [1, 120–125].

Earlier reviews concerning the preparation of labelled gibberellins had been already appeared in the literature from J. R. Hanson [126], J. MacMillan [127] and N. Takahashi/I. Yamaguchi [128].

4.1 Carbon Labelling

4.1.1 Biochemical Methods

[1,7,12,18-^{14}C]GA$_3$ has been prepared by a biochemical method from [2-^{14}C]DL-mevalonate (**124**) with the fungus *Giberella fujikuroi* [129, 130]. Similarly, the ^{14}C-labelled compounds GA$_1$, GA$_4$, GA$_7$, GA$_9$ and GA$_{13}$ have been synthesized [131]. Also [U-^{14}C]GA$_3$ was obtained from [1,2-^{14}C] NaAc with the same fungus [132]. This labelled compound was obtained with a specific activity of only 74 MBq/mmol. Starting from [1-^{13}C]acetate, Yamaguchi et al. [89] synthesized ^{13}C-la-labelled GA$_1$ and GA$_3$ with *Gibberella fujikuroi*.

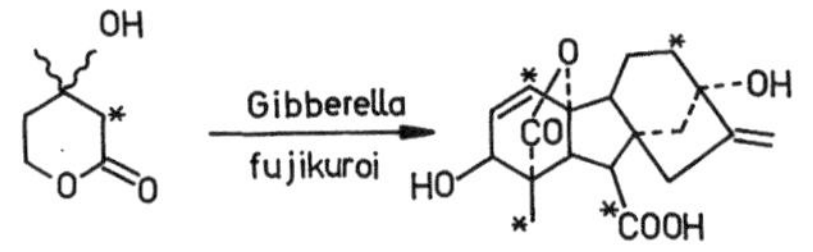

4.1.2 Chemical-Biochemical Methods

In special cases the chemical-biochemical method can be a better way for preparation of labelled gibberellins. Hanson and Hawker [133] prepared [17-^{14}C]-GA$_3$ with a specific activity of 207 MBq/mmol starting from GA$_{12}$-7-alcohol (**125**) via cleavage of the terminal methylene group to the 16-*nor*ketone and Wittig reolefination with $Ph_3P{=}^{14}CH_2$ to give **126** which was further transformed with *Gibberella fujikuroi*. Similarly, the labelled gibberellins [17-^{14}C]GA$_3$, [17-^{14}C]GA$_4$ and [17-^{14}C]GA$_7$ were obtained [134].

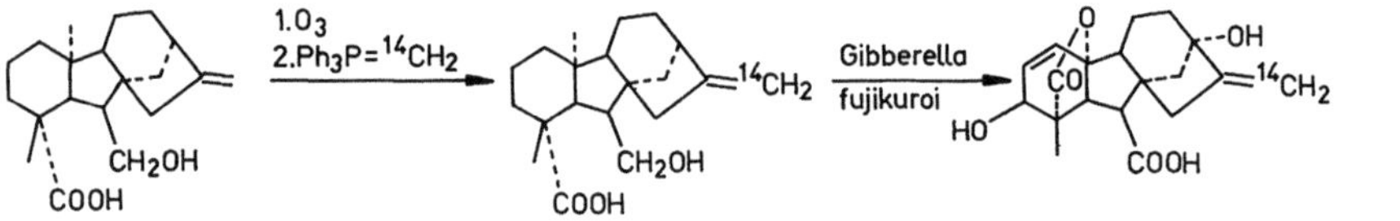

4.1.3 Chemical Methods

In general, the chemical methods afford labelled gibberellins with the highest specific activities. Starting from corresponding 16-oxo-17-nor gibberellin derivatives, the following [17-^{14}C] labelled gibberellins were obtained by Wittig reaction:

[17-^{14}C]GA$_{12}$ (spec. act. 1,5 MBq/mmol), [17-^{14}C]GA$_{13}$ (spec. act. 0,2 MBq/mmol), [17-^{14}C]GA$_{14}$ (spec. act. 0,18 MBq/mmol) [135], [17-^{14}C]GA$_9$ (spec. act. 0,15 MBq/mmol) [136].

As shown in the following reaction sequence [17-^{13}C]GA$_{29}$ was obtained with 0.97 atoms ^{13}C per molecule [95]: Starting from the 3-epimeric GA$_1$ derivatives **21** and **50** fission of the terminal methylene group with $OsO_4/NaIO_4$ and treatment

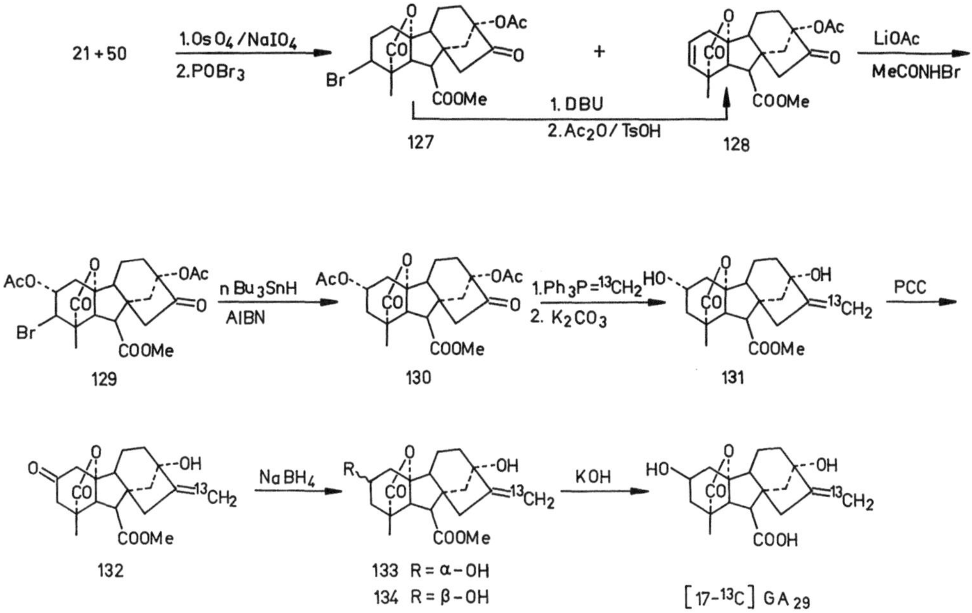

of the obtained 16-keto-17-nor-compounds with $POBr_3$ gave the 3β-bromo compound **127** as well as the elimination product **128**. Compound **128** was also obtained by treatment of **127** with DBU. Functionalization of the Δ^2 double bond with LiOAc/MeCONHBr led to the 2α-acetoxy-3β-bromo compound **129**. Subsequent stannane reduction to **130** followed by Wittig reolefination with $Ph_3P={}^{13}CH_2$ and alkaline hydrolysis gave the 17-labelled diol **131**. To obtaine the native 2β-configuration, **131** was oxidized with PCC to **132** which afforded upon borohydride reduction both 2-epimers **133** and **134** in an 1:1 ratio. Hydrolysis of the separated **134** led to the desired [17-^{13}C] GA_{29}.

4.2 Hydrogen Labelling

4.2.1 Labelling with $^3H_2$ or $^2H_2$

[1,2(n)-^{3}H]GA_1 and [1,2(n)-^{3}H]GA_4 can be obtained commercially with a specific activity of 0.92–1.85 TBq/mmol. Both gibberellins have been prepared by catalytic tritiation of the Δ^1 double bond in GA_3 or GA_7. This labelling method has been often used in the literature, especially for [1,2-$^3H_2$]GA_1 [84, 137–141]. [^{3}H]GA_3 was obtained by exchange with a specific activity of 481 GBq/mmol [138]. However, MacMillan and Willis [91] found a better method for specific labelling of the Δ^1 double bond. Thus, GA_3 methyl ester-13-acetate, prepared in three steps from GA_3, was epoxidized to **135** followed by deuteriogenation to give **136**. Subsequent deoxygenation followed by step-wise hydrolysis of the ester functions led to [1,2-$^2H_2$]GA_1 containing 1.56 atoms of deuterium per molecule.

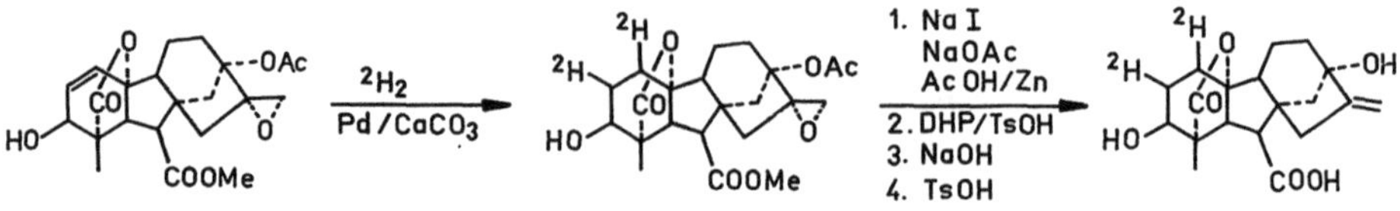

135 136 $[1,2\text{-}^2H_2]$ GA_1

Durley and Pharis [142] obtained directly $[1,2\text{-}^3H_2]GA_4$ by tritiation of the Δ^1 double bond of GA_7 (spec. act. 69.2 GBq/mmol). Similar, $[1,2\text{-}^2H_2]GA_4$ methyl ester was formed containing 1.82 atoms of deuterium per molecule [143]. On the other hand, tritiation of the Δ^2 double bond in **137** gave $[2,3\text{-}^3H_2]GA_9$ (spec. act. 1.74 TBq/mmol) [144].

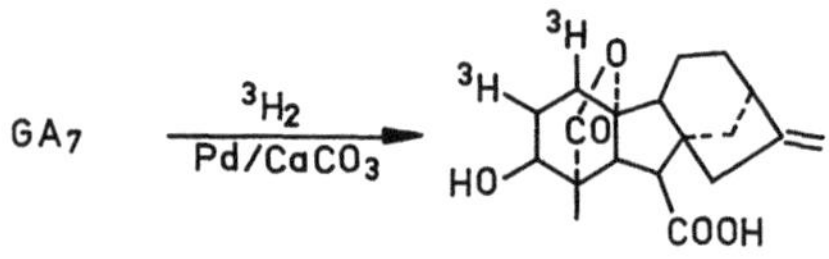

$[1,2\text{-}^3H_2]$ GA_4

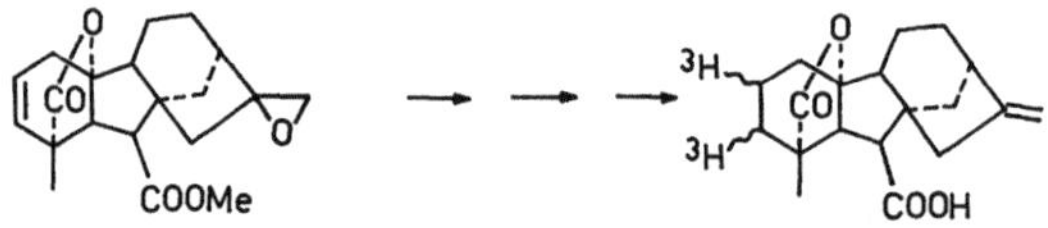

137 $[2,3\text{-}^3H_2]$ GA_9

Catalytic tritiation of the epoxide **48** (see Sect. 3.10) led to $[2,3\text{-}^3H_2]GA_{20}$ (spec. act. 74 GBq/mmol) [92, 145, 146].

However, detailed ^3H-NMR studies of this reaction showed the additional incorporation of tritium at 1β position [147, 148]. The resulting $[1\beta,2\beta,3\beta\text{-}^3H_3]GA_{20}$ had a specific activity of 2.04 TBq/mmol. In analogy, the $[1\beta,2\beta,3\beta\text{-}^2H_3]GA_{20}$ was obtained with an incorporation rate of 24% $^2H_3$, 41% $^2H_2$ and 28% $^2H_1$.

Analogous labelling of **20** yielded after hydrolysis of the ester group $[1\text{-}^3H]\text{-}GA_5$ with a specific activity of 196.1 GBq/mmol [92, 146]. Starting from tritiated GA_1-derivatives $[1,2\text{-}^3H_2]GA_5$ was obtained by H_2O elimination [149, 150]. On the other hand, $[U\text{-}^3H]GA_3$ (spec. act. 74 MBq/mmol and 600 MBq/mmol) has been prepared using either the Wilzbach method or by tritium exchange via heterogeneous catalysis [151–153].

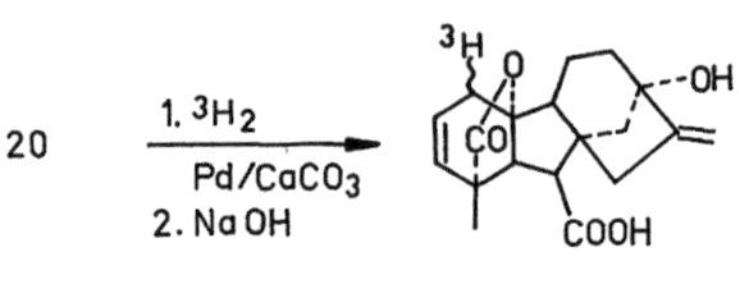

$[1\text{-}^3H]$ GA_5

4.2.2 Labelling with ^3H_2O or ^2H_2O

The labelling of gibberellins is also possible using ^3H_2O or ^2H_2O. The partial synthesis of [6-^{3}H]GA_3 [154, 155] (specific activity 833 GBq/mmol) has been realized starting from GA_3 via the intermediates diacetyl-GA_3, tetraacetyl-GA_3 anhydride, diacetyl-GA_3-7-alcohol (**138**) and GA_3-7-aldehyde (**139**) [156] which was silylated and enolized to give **140**. Addition of tritiated water (prepared by oxidation of carrier-free tritium-gas) gave [6-^{3}H]GA_3-7-aldehyde (**141**) which was acetylated, oxidized and deacetylated to [6-^{3}H]GA_3. This method allowed also the tritium labelling of GA_1, GA_4, GA_5, GA_7 and GA_9 [157].

[6-^{2}H]GA_3 (>0.9 atoms ^{2}H per molecule) was obtained analogously using ^2H_2O [158].

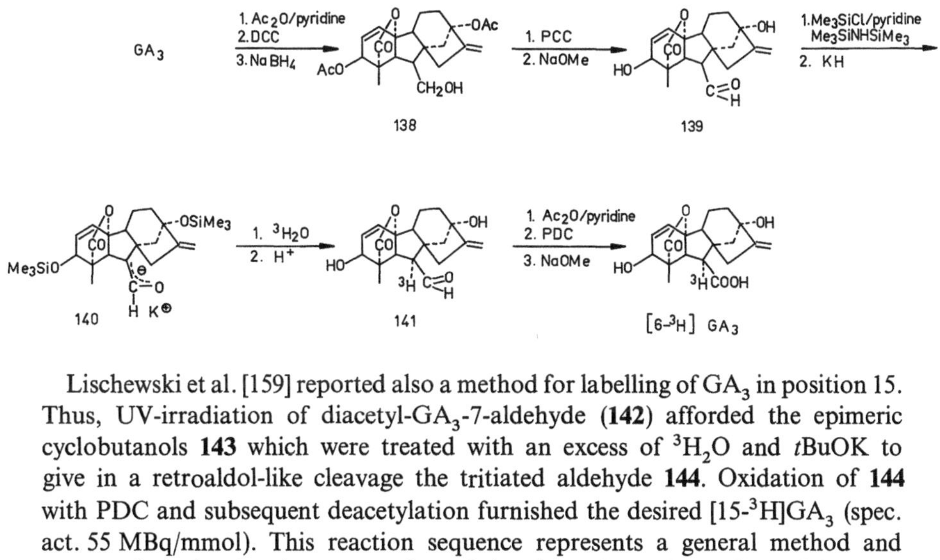

Lischewski et al. [159] reported also a method for labelling of GA_3 in position 15. Thus, UV-irradiation of diacetyl-GA_3-7-aldehyde (**142**) afforded the epimeric cyclobutanols **143** which were treated with an excess of ^3H_2O and *t*BuOK to give in a retroaldol-like cleavage the tritiated aldehyde **144**. Oxidation of **144** with PDC and subsequent deacetylation furnished the desired [15-^{3}H]GA_3 (spec. act. 55 MBq/mmol). This reaction sequence represents a general method and was applied also for the preparation of the GA_1, GA_4, GA_5, GA_7, GA_9 [159] and [15-^{2}H]GA_3 [160].

[6-^{3}H]GA_{12} (spec. act. 27.4 MBq/mmol) has been synthesized by base-catalyzed exchange reaction of the corresponding aldehyde **145** to obtain **146** and subsequent

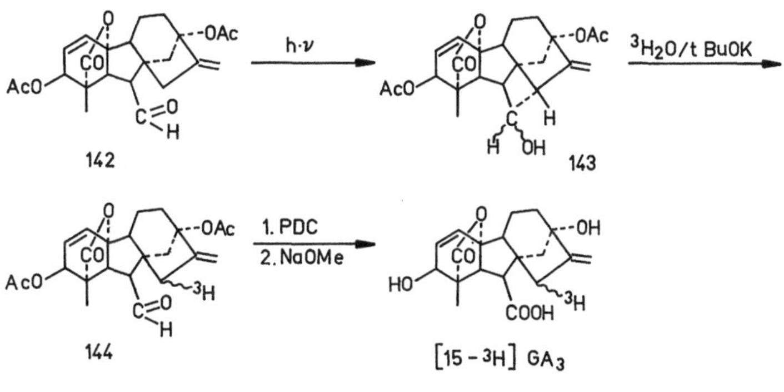

oxidation. An analogous sequence using ^2H_2O led to [6-^{2}H]GA_{12} (0.90 atoms ^{2}H per molecule) [161]. On the other hand, Hanson et al. [162] synthesized [6-^{3}H] GA_{12}-7-aldehyde (**146**) (spec. act. 0.28 MBq/mmol) by ring contraction of a kaurene derivative.

GA_{14}-aldehyde was also labelled by exchange reaction [163]. In this case, the resulting [6-^{3}H]GA_{14}-aldehyde (spec. act. 11 MBq/mmol) was reduced to the [6-^{3}H]GA_{14}-alcohol (spec. act. 7.4 MBq/mmol) [163].Yamaguchi et al. [89] obtained [1-^{2}H]GA_1 (0.2 atoms ^{2}H per molecule) by acid catalyzed lactonization of 8 in presence of ^2H_2O and subsequent Bartlett demethylation.

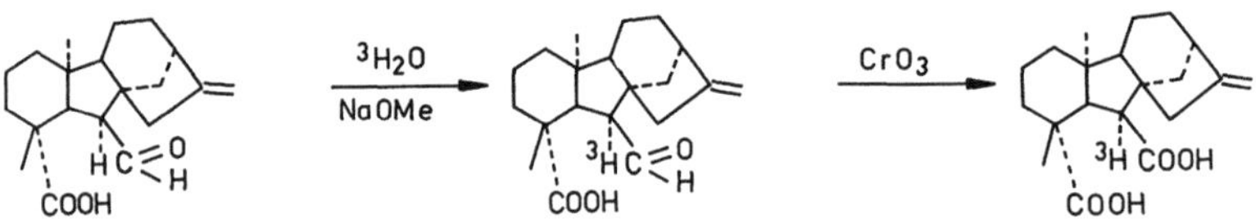

145 146 [6-^{3}H] GA_{12}

Alternatively, the ylide $Ph_3P{=}C^3H_2$ can be used for labelling. The salt of this compound was prepared from $[Ph_3P^{\oplus}\text{-}CH_3]Br^{\ominus}$ by treatment with ^3H_2O under alkaline conditions. Synthesis of [15,17-$^3H_4$]GA_9 was achieved via Wittig reaction of 16-oxo-17-*nor*-GA_9 spec. act. 2.3 GBq/mmol and 1.07 GBq/mmol) [164, 165]. Bearder et al. [165] proved the localization of the labelling in [15,17-$^3H_4$]GA_9 by ozonolysis to the corresponding *nor*ketone (remained activity 41%) and subsequent exchange reaction with *n*BuOH/*t*BuOK (remained activity

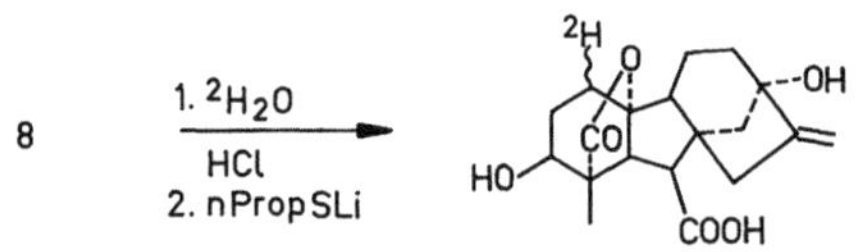

[1-^{2}H] GA_1

0.06%). [17-^{3}H]GA_{14} was formed using the same method with a specific activity of 2.25 GBq/mmol [166]. Similarly, the synthesis of [17-^{13}C] [15,17-^{3}H]GA_{29} (0.91 atoms ^{13}C per molecule; spec. act. 1.1 GBq/mmol) was achieved by Wittig olefination of *nor*ketone **130** with $Ph_3P{=}^{13}C^3H_2$ [95]. Analogous reaction using o(13)-acetyl-16-oxo-17-*nor*-GA_{20} methyl ester and $Ph_3P{=}^{13}C^3H_2$ gave after alkaline hydrolysis [17-^{13}C] [15,17-^{3}H]GA_{20} (0.88 atoms ^{13}C per molecule; 1.26 GBq/mmol) [57]. As an alternative to the Wittig olefination [17-$^2H_2$]GA_4 (containing 0.99 atoms ^{2}H per molecule) can be obtained by reaction of 16-oxo-17-*nor*-GA_4 with $Zn/C^2H_2Br_2/TiCl_4$ [96].

4.2.3 Labelling with Tritiides and Deuterides

As starting material for labelling of gibberellins in position 1,2 and 3, the methyl ester of the enone **13** was reduced with NaB^2H_4/LiBr in dimethoxyethane to give both labelled 3-epimeric GA_1 derivatives **21** and **50** besides the Δ^1-unsaturated

3-epimers (see Sect. 3.1). Dehydration of 21 with $POCl_3$/pyridine and hydrolysis gave the desired [1β,3-$^2H_2$]GA_5 (1.70 atoms ^2H per molecule) [109]. The 3α-hydroxy compound **50** was transformed with $POCl_3$/pyridine to the corresponding 3β-chloro compound which gave after stannane reaction and hydrolysis [1β,3α-$^2H_2$]GA_{20} (1.86 atoms ^2H per molecule). Analogous sequences using NaB^2H_4 in the labelling step gave [1β,3-$^3H_2$]GA_5 and [1β,3α-$^3H_2$]GA_{20} [109]. Similar, [2α-^2H]GA_{20} and [3α-^2H]GA_{20} were prepared [147].

In an improved method, Duri and Hanson [168] used NaB^2H_4/CuCl in MeOH for the 1β, 3β deuteriogenation of the methyl ester of the enone **13**. Reduction with $NaBH_4$/CuCl in MeO^2H effected a 2α:2β deuteriogenation in a 0.35:1 ratio. Alternatively, this reaction can be achieved also with NaB^2H_4/CuCl in MeO^2H with deuteriogenation in positions 1β,2α,2β and 3β.

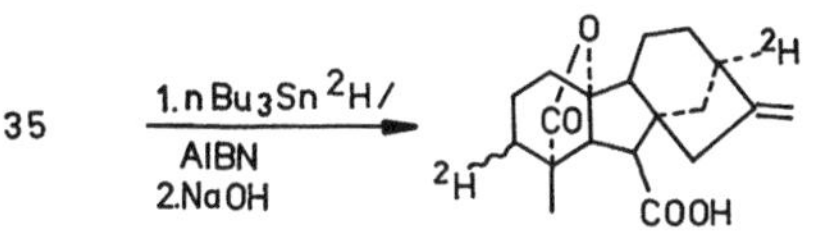

[3,13-$^2H_2$] GA_9

3-*epi*-GA_1 methyl ester (**34**) has been used as starting material for double labelling of GA_9 in the positions 3 and 13. Similar as describes for synthesis of unlabelled GA_9 (see Sect. 3.9) reduction of the dichloro intermediate **35** with $n$$Bu_3Sn^2H$ and alkaline hydrolysis gave [3,13-$^2H_2$]GA_9 [100]. As described above, the 3-dehydro-GA_7 methyl ester can be used for $NaBH_4$ reduction [99].

Beale and MacMillan [169] published a procedure for preparation of labelled GA_{29} and GA_{51}. For synthesis of [2α-^3H]GA_{29} spec. act. 2.2 GBq/mmol) in the manner described in Sect. 4.1.3, the **132** corresponding unlabelled 2-ketone was reduced with NaB^3H_4 to give the both 2-epimeric-hydroxy compounds and the separated 2β-epimer was demethylated. Analogous reduction with NaB^2H_4 afforded [2α-^2H]GA_{29} (containing 0.97 atoms ^2H per molecule) [169].

In the same publication, the synthesis of double labelled [2α-^2H] [15,17-$^3H_4$]GA_{51} was reported. Starting from the *nor*ketone **147**, the terminal methylen group was reconstructed by Wittig reolefination with $Ph_3P{=}C^3H_2$ to give **148**. Subsequently, the deuterium labelling was introduced in position 2α via CrO_3 oxidation followed by NaB^2H_4 reduction and hydrolysis.

Beale et al. [109] published the stereospecific deuteriogenation of GA_9 in position 3: Thus, the 3-oxo ester **149** was reduced with NaB^2H_4 to give preferred the 3β-deuterated alcohol **150**. Transformation to the 3β-chloro compound **151**, stannane reduction and hydrolysis afforded [3α-^2H]GA_9 (0.92 atoms ^2H per

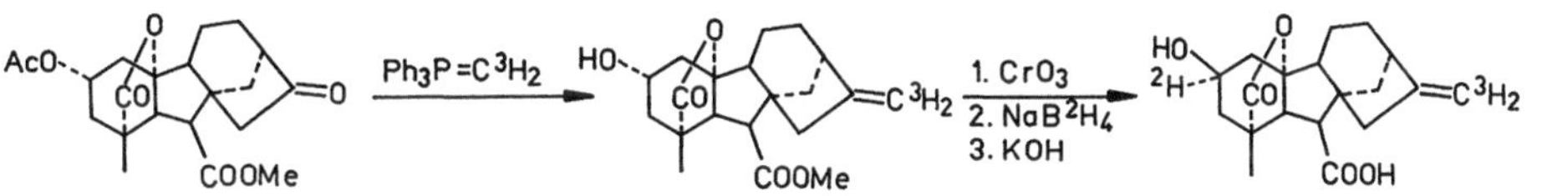

147 148 [2α -^2H][15,17-$^3H_4$] GA_{51}

molecule). For the synthesis of [3β-^{2}H]GA$_9$, (0.96 atoms ^{2}H per molecule), the dichloro compound **37** (see Sect. 3.9) was reduced with nBu$_3$Sn^{2}H and demethylated. Similarly, 2ξ deuterated GA$_9$ (0.99 atoms ^{2}H per molecule) and [2ξ-^{2}H]GA$_4$ (0.84 atoms ^{2}H per molecule) were synthesized [99].

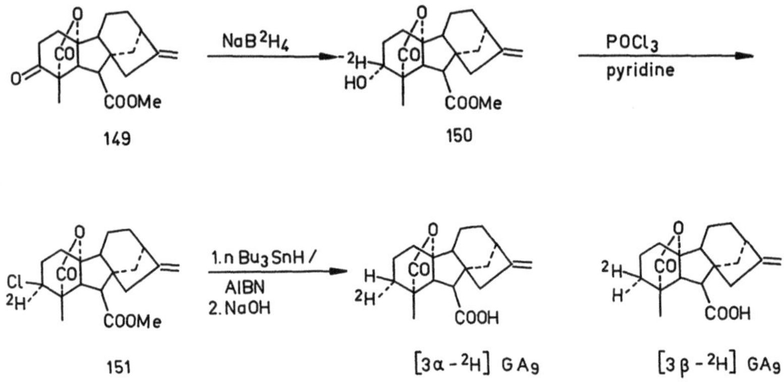

Stereospecific introduction of ^{2}H in position 2α of GA$_4$ and GA$_9$ can be achieved according following sequences [99]: Starting from the 3-ketone **149**, metallation with lithium diisopropylamide followed by transformation to the corresponding 2β-iodo-ketone and reduction with nBu$_3$Sn^{2}H gave the 2α-deuterated 3-ketone **152**. Stereospecific reduction of **152** with aluminium foil/mercuric chloride to the 3β-hydroxy compound **153** followed by Bartlett demethylation gave [2α-^{2}H]GA$_4$ (0.90 atoms ^{2}H per molecule). On the other hand, **152** was transformed to the 3β-chloro compound **154**. Subsequent removal of the chloro function and hydrolysis gave [2α-^{2}H]GA$_9$ (0.80 atoms ^{2}H per molecule).

The first labelling of a gibberellin in position 18 was described by Down et al. [170]. Thus, the starting kaurene alcohol **155** was transformed with Ph$_3$P/CCl$_4$ to the 18-chloro compound **156**. Subsequent reduction with nBu$_3$Sn^{2}H led to the 18-deuteriated compound **157**, which was oxidized in position 7 and followed by borohydride reduction to the 7α-alcohol and esterification with p-bromo-benzoyl-

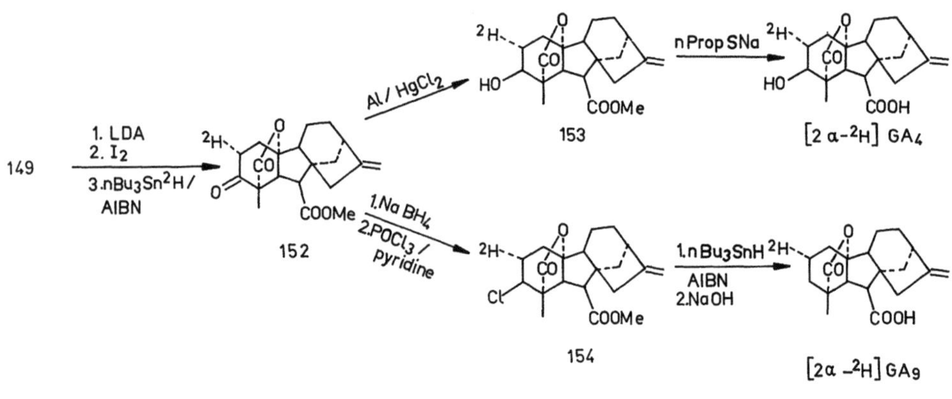

sulfonylchloride to give **158**. Alkaline catalyzed ring contraction led to [18-^{2}H]GA$_{12}$ aldehyde (0.88 atoms ^{2}H per molecule).

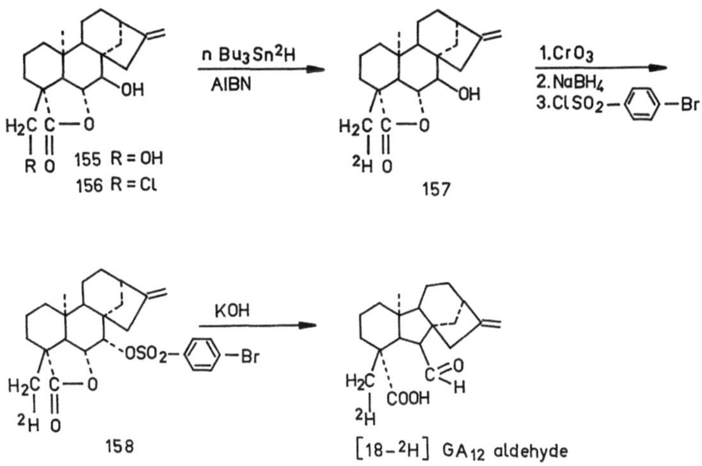

5 Partial Synthesis of Structurally Modified Gibberellins

In connection with the structural elucidation, derivatization and intercorrelation of gibberellins, as early as 1965, nearly 800 gibberellin compounds had been described [171]. Later for studies on structure-activity relationships, mode of action and receptor considerations [172–174] more special synthetic programms were carried out by several groups. These efforts also included systematical studies of photochemical transformations of gibberellins opening up further original pathways to new structural types [175]. In the following, the recent developments in this field will be discussed.

5.1 6-Modified Gibberellins

To verify the essential role of the 6β-carboxylic function [176] for a high biological activity as well as for transport and putative receptor interaction a series of 6-modified GA$_3$ analogs was synthesized leaving intact all other structural features of this hormone molecule. Thus, GA$_3$ was transformed in a four-step sequence via diacetate **138** to GA$_3$-7-alcohol (**159**) and in 5 steps to GA$_3$-7-aldehyde (**139**) [156], the latter one used also as intermediate for labelling at C-6 and 15 (see Sect. 4.2.2). Compound **139** via enolization of its trimethylsilyl derivative and deprotection gave an equilibrium of **139** and its 6α-epimer **160** separable by chromatographic methods. Acetylation of **160** followed by PDC-oxidation and hydrolysis afforded 6-*epi*-GA$_3$ (**161**) [177]. The fully deoxygenated GA$_3$ analog **162** was prepared from diacetyl alcohol **138** via the corresponding iodomethyl compound, its reduction with triethyltinhydride and deacetylation. The synthesis of the homologue **163** was achieved from GA_3 in five steps using photo Wolff rearrangement of the diazoketone of diacetyl GA_3 as the key reaction. Another pathway envolved the preparation of the iodomethyl intermediate from diacetyl alcohol **138** and its subsequent reaction with disodium tetracarbonyl-(II)-ferrate to give a *homo*aldehyde which was further oxidized and deprotected to **163** [178]. 7-*nor* GA_3 (**164**) is available from GA_3 by lead tetraacetate decarboxylation of its diacetate in benzene and hydrolysis [179]. For preparation of the nitrile **165** GA_3 was transformed via reaction of its anhydride with liquid ammonia to GA_3-7-amide. Subsequent acetylation, dehydration to the nitril and hydrolysis gave compound **165** [180].

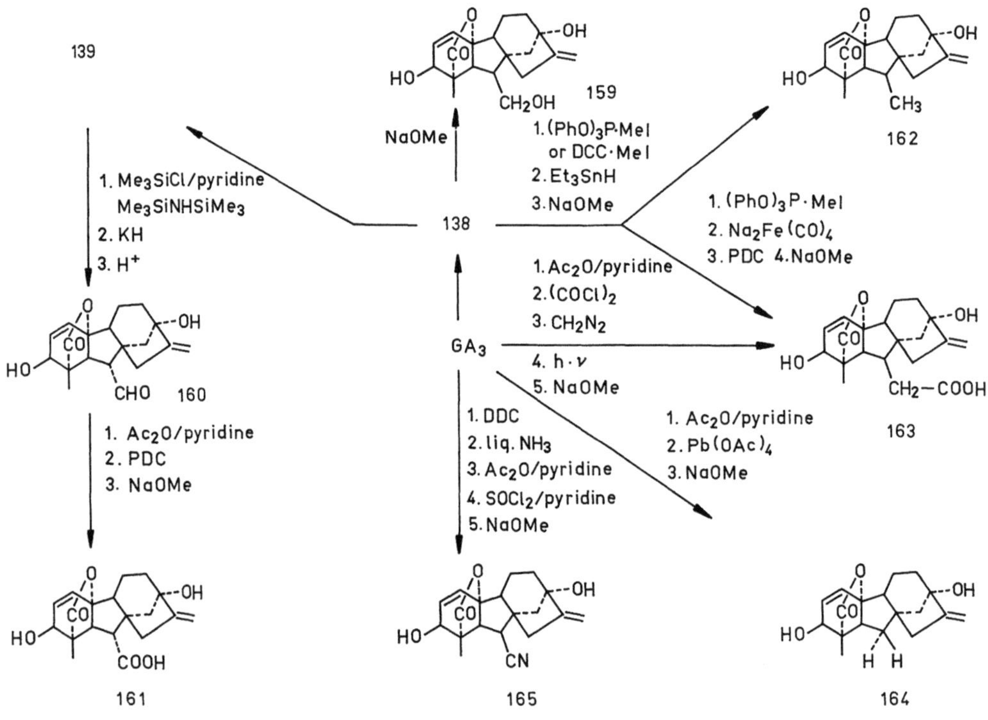

The series of the so obtained 6-modified GA_3 analogs has been the subject of detailed studies of their biological properties [181]. Interestingly, 7-*nor* GA_3 (**164**) showed a remarkable phytohormone activity in selected bioassay systems and seems to be the first active gibberellin which is selectively transported and distributed within the plan [75]. On the other hand some *ent*-7-*nor*-gibberella-5,10-dienes, synthesized from the fungal metabolite fujenal [182], were shown to act as plant growth inhibitors and as inhibitors of GA_3 biosynthesis in *Gibberella fujikuroi* [183]. Starting from GA_{13} the corresponding 7-alcohol and 7-aldehyde have been prepared via 19,20-anhydrides [184]. Also the methyl esters of 6-*epi*-GA_{13} and 6-*epi*-GA_{25} can be derived from GA_{13} [115].

5.2 Halogenated Gibberellins

Animal hormone research has shown, especially in the steroid field, that introduction of a halogen is a successful strategy to increase or modify bioactivity. In particular, replacement of hydrogen with fluorine causes only a small increase in size but results in a remarkable enhancement of electronegativity and hydrogen bonding potential. For that reason also in the field of gibberellin phytohormones, considerable efforts have been invested into synthesis of fluorinated and other halogenated analogs.

The first compounds of this structure were prepared in 1972 by fermentation of the fluoraldehyde **166**, available from 7,18-dihydroxy kaurenolide, with *Gibberella fujikuroi* to give 18-fluoro GA_3 (**168**) (as a mixture with 70% GA_3) and 18-fluoro GA_9 (**169**). Jones oxidation of **166** afforded 18-fluoro GA_{12} (**167**) [186, 187]. Compounds **167** and **169** exhibited, in several bioassays, similar activities to the non-fluorinated parent gibberellins whereas 18-fluoro GA_3 (**168**) seems to be a competitive GA_3 inhibitor [188]. Other fluorinated gibberellin derivatives were also formed from fluorinated *ent*-kaurene precursors by the mentioned fermentation procedure [189, 190].

In chemical experiments direct fluorination of GA_3 phenacyl ester (**170**) with 1,2 mol equiv. 2-chloro-*N*,*N*-diethyl-1,1,2-trifluoro-ethylamine (fluoroamine) followed by reductive de-esterification led to the 3*β*-fluorinated GA_3 analog **171** and 1*β*-fluoro GA_5 (**173**) [191]. Fluorination of **170** with an excess of the fluoroamine caused additional substitution of the tertiary 13-hydroxyl to give after de-esterification both difluoro gibberellins **172** and **174**. Under same conditions the mixture of 16-epimeric dihydro-GA_1 methyl ester (**175**) yielded the Δ^2-13-fluoro ester **176** which was hydrolyzed to 13-fluoro-dihydro GA_5 (**177**) as well as hydrogenated and hydrolyzed to give 13-fluoro-dihydro GA_9 (**178**) [192]. More conveniently the fluoro olefin **176** can be prepared by treatment of **175** with diethylaminosulphur trifluoride (DAST). Compound **178** was transformed by diborane reduction into its 7-alcohol **179** which upon fluorination with DAST gave the difluoro derivative **180** [193]. In interaction studies with the unfluorinated com-

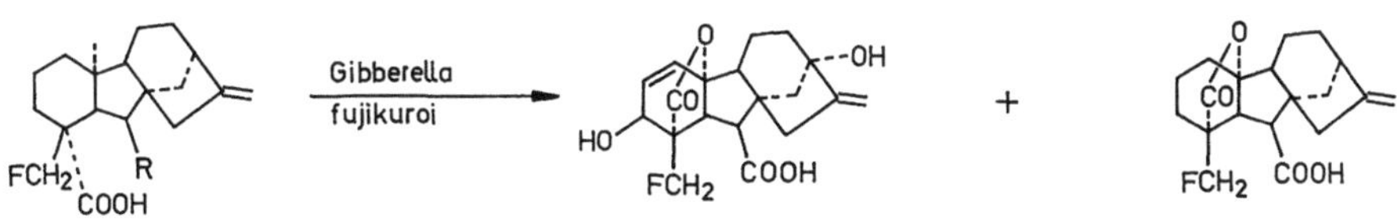

166 R = CHO
167 R = COOH

168

169

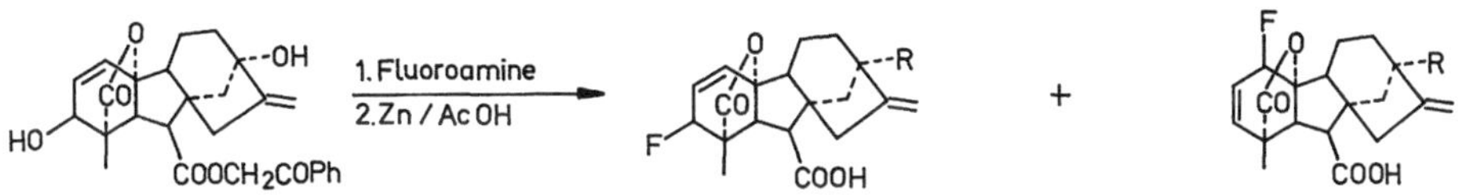

170

171 R = OH
172 R = F

173 R = OH
174 R = F

pounds the fluorogibberellins **172**, **174**, **177** and **178** generally showed competitive inhibiting activity [194].

By treatment of GA_{40} methyl ester (**83**) with fluoroamine and hydrolysis 2β-fluoro GA_9 (**181**) (containing 10% of its Δ^{15}-isomer) is available [195]. Its 2α-epimer **186** was synthesized from the GA_4-*nor*ketone **182** via the Δ^2-phenacyl ester **183** which upon addition of BrF gave the 2α-fluoro-3β-bromo derivative **184**. Subsequent reductive hydrolysis and elimination of bromine afforded **185** which was transformed by Wittig reolefination to the required 2α-fluoro GA_9 (**186**).

3β-Fluoro GA_9 (**188**) and 3β-chloro GA_9 (**189**) are formed in an 1:2,5 ratio by reaction of 3-*epi*-GA_4 phenacyl ester (**187**) with fluoroamine and reductive hydrolysis. The fluoro analogs **186** and **188** as well as the 3β-chloro compound **189** showed similar bioactivity than the parent hormone GA_9 [196].

For conversion of hydroxygibberellins to chlorides the fluoramine/LiCl reagent was found to be suitable. In such a manner from the benzyl ester of GA_3 via a corresponding Δ^2-unsaturated 1β,13-dichloro ester and hydrogenation over rhodium/alumina among other products 13-chloro-dihydro GA_9 (**190**) and 1β,13-dichloro-dihydro GA_9 (**191**) were obtained as 13-epimers. Direct chlorination of GA_3 led mainly to a trichloro-bis-compound. Pure (13*S*)-13-chloro-dihydro GA_5 (**193**) could be synthesized by stereospecific reduction of GA_3 with diimide to afford (13*S*)-tetrahydro GA_3 (**192**) followed by reaction with fluoroamine/LiCl. The three chlorogibberellins **190**–**191** and **193** were found to be more active than dihydro GA_5 but less active than GA_9 and the 13-monohalogenated analogs **190** and **193** were more active than **191** in the dwarf rice bioassay [197].

1β-Iodo GA_1 (**194**) and 1β-iodo GA_4 arise as intermediates in the relactonization of corresponding $\Delta^{1(10)}$-7,19-dioic acids with $I_2/NaHCO_3$ (see, 3.1. and 3.3.) [86, 198]. In a modified procedure recently 1β-bromo GA_1 (**195**) and 1β-

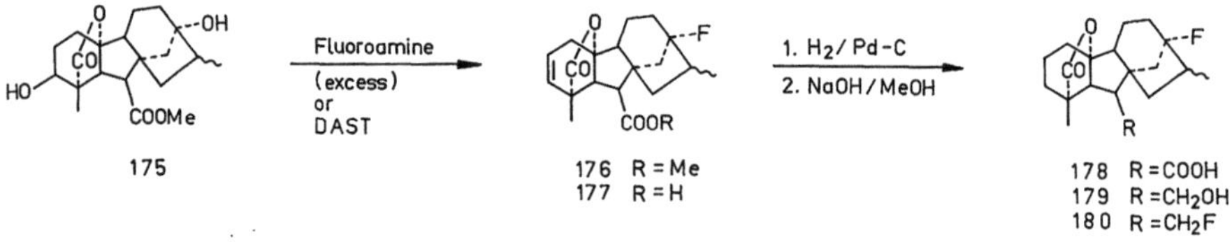

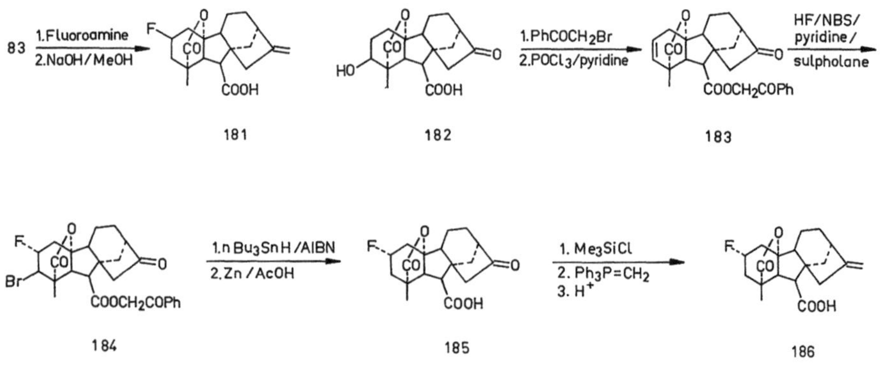

chloro GA_1 (**196**) were prepared from GA_3 via hydrogenation to the 8 corresponding $\Delta^{1(10)}$-dioic acid and relactonization using *N*-halogeno succinimide/$NaHCO_3$ as source for the hypohalogenous acid [199]. The three 1β-halogenated GA_1 analogs **194–196** showed similar bioactivity than the parent hormone GA_1 [75, 198].

Upon treatment of GA_3 with $SOCl_2$/pyridine at —20 °C under inversion at C-3 the 3α-chloro gibberellin **197** is available [200]. Reaction of GA_3 with phenyliodo chloride under UV-irradiation afforded 16*S*,17-dichloro GA_3 (**198**) in moderate yield [201].

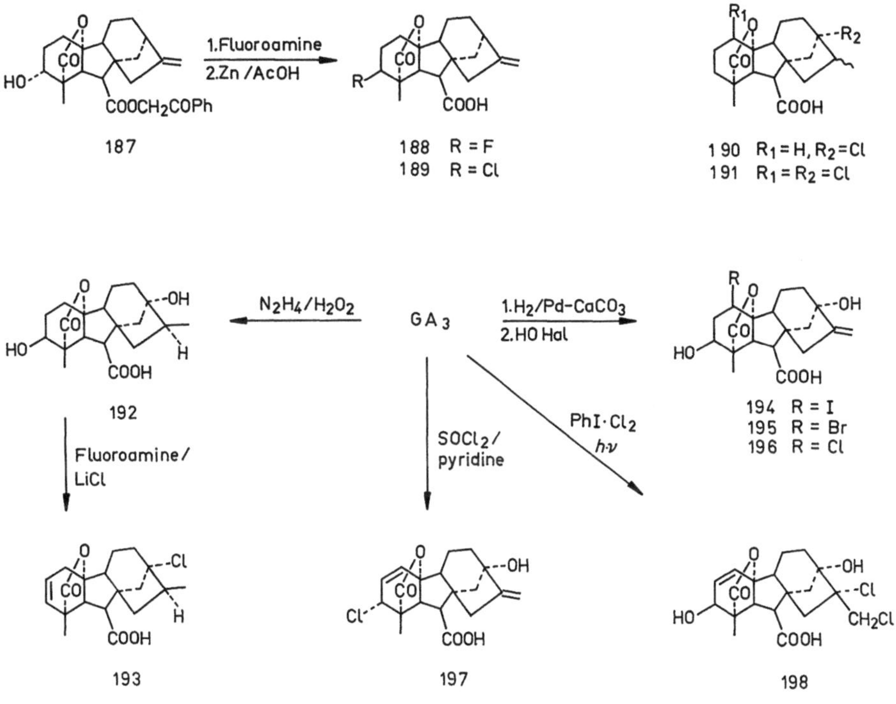

5.3 Alkylated Gibberellins

Published data show that introduction of substituents at C-2 can increase the biological activity by blocking the metabolic deactivation of gibberellins via 2β-hydroxylation [202]. Based on this concept, detailed investigations were directed towards the synthesis of gibberellins alkylated at this position.

The preparation of 2β-methyl and 2,2-dimethyl analogs of the GA_4 and GA_9 series was first described by Beale and MacMillan [195]. Thus, exhaustive methylation of 3-dehydro-GA_4 methyl ester (**149**) with methyl iodide/potassium *t*-butoxide gave besides a *seco*-ring A acid the desired 2,2-dimethylketone (**199**) which was reduced with borohydride to afford the 3-epimeric alcohols **200** and **203**, respectively. To avoid epimerization at C-3, compound **200** was transformed to the 3β-

tetrahydropyranyl ether, whiche gave upon alkaline hydrolysis and acid deprotection 2,2-dimethyl GA_4 (**201**). The 3α-alcohol **203** was converted via the 3β-chloro compound, stannane reduction and hydrolysis to 2,2-dimethyl GA_9 (**204**). By incubation with *Gibberella fujikuroi* mutant B41a compound **201** could be hydroxylated preparatively to 2,2-dimethyl GA_1 (**202**). Upon alkylation of **149** with one equivalent MeI among other products the 2β-monoalkylated ketone **205** was available and was transformed in analogous sequences to 2β-methyl GA_4 (**206**) as well as to 2β-methyl GA_9 (**207**).

In an improved synthesis for 2,2-dimethyl GA_4 (**201**) instead of methyl ester **149** the corresponding tri-*n*-butyltin ester was used. Its methylation with MeI/Li-di-isopropamide followed by reduction of the 3-keto function with sodium isopropoxy borohydride led to **201** and its 3α-epimer [203].

The synthesis of 2β-alkylated GA_1 compounds from GA_3 was realized as follows [204]: Upon direct alkylation of GA_3 with lithium dialkylcuprates *anti*-S_N2'-type substitution [205] with displacement of the allylic lactone function occurred. The formed 2β-alkyl-$\Delta^{1(10)}$-dioic acids **208**–**211** were recyclized via iodolactonization to **212**–**215** which gave via reductive de-iodination of their 7-stannyl esters with nBu_3SnH the desired series of the 2β-alkylated GA_1 analogs **216**–**219**. As proved by GC-MS in all cases the 2β-epimers were preferred, however the amount of the 2α-epimer increased with prolonged alkyl chain. The use of GA_3 methyl ester (**6**) instead of GA_3 gave higher yields but required additional Bartlett demethylation in the last step. As a reagent for a stereoselective *syn*-S_N2'-type displacement of the allylic lactone function of **6** an alkyl copper-magnesium complex was found to be suitable [206]. The so available 2α-alkyl-$\Delta^{1(10)}$-19-oic acids **220**–**222** were relactonized and demethylated to 2α-methyl-, ethyl- and *n*-propyl GA_1 (**223**–**225**) opening a general route to 2α-alkylated gibberellins. By 13-deoxy-

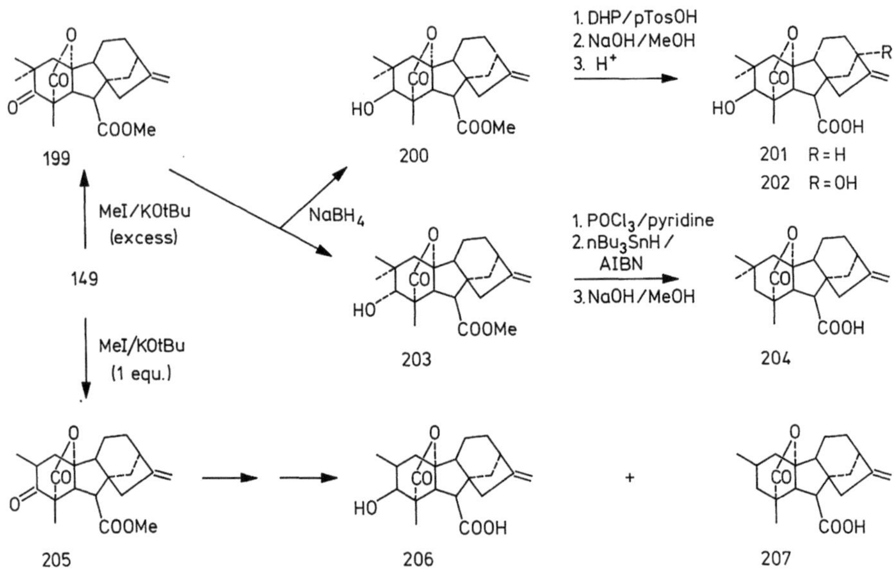

genation of **224** via stannane reduction of the corresponding 13-oxalyl ester 2α-ethyl GA_4 (**226**) was prepared.

Recently, pathways to GA_4 analogs with branched 2β-alkyl substituents have also been published [207]. As key intermediate the synthon **227**, available from **149** by treatment with lithium 2,4,6-tri-*t*-butylphenoxide/CS_2 and MeI, was used. For example, reaction of **227** with 4 mol equiv. lithium dimethylcuprate gave preferentially the 2β-*t*-butyl ketone, which was reduced with $NaBH_4$ to a 4:1 ratio of the 3-epimeric alcohols. Demethylation of the main β-epimer afforded 2β-*t*-butyl-GA_4 (**228**). Modified sequences led to 2β-isopropylidene- and 2β-isopropyl-GA_4 as well as to the spiro compound **229**.

For preparation of 1-methyl- and 1α,2α-cyclopropyl analogs gibberellin pyrazolins, available from corresponding enones with diazomethane, were found to be suitable starting compounds. Thus, 3-dehydro-GA_7 methyl ester (**116**) yielded the pyrazoline ketone **230** which was thermolyzed to give 1-methyl enone **231**. Subsequent reduction with $NaBH_4$/CuCl afforded mainly the 1α-methyl compound and upon demethylation the desired 1α-methyl GA_4 (**232**). Compound **232** was deoxygenated as described (see Sect. 3.9) to give 1α-methyl GA_9 (**233**). For preparation of 1β-methyl GA_4 (**237**) the enone **231** was reduced with $n Bu_3SnH$ in presence of tetrakistriphenylphosphinepalladium(O) to yield a 1:1 mixture of 1-epimeric methyl ketones **235** and **236** which afforded upon borohydride reduction and alkaline hydrolysis an inseparable 7:3 mixture of 1α- and 1β-methyl GA_4 **232** and **237**, respectively. Upon reoxidation to the 3-ketones, HPLC-se-

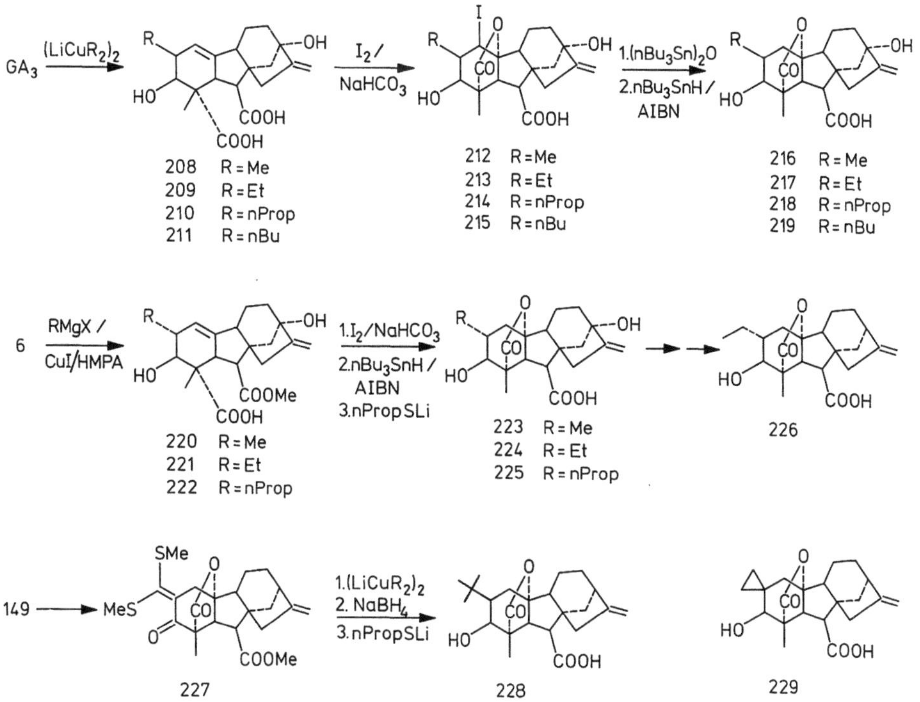

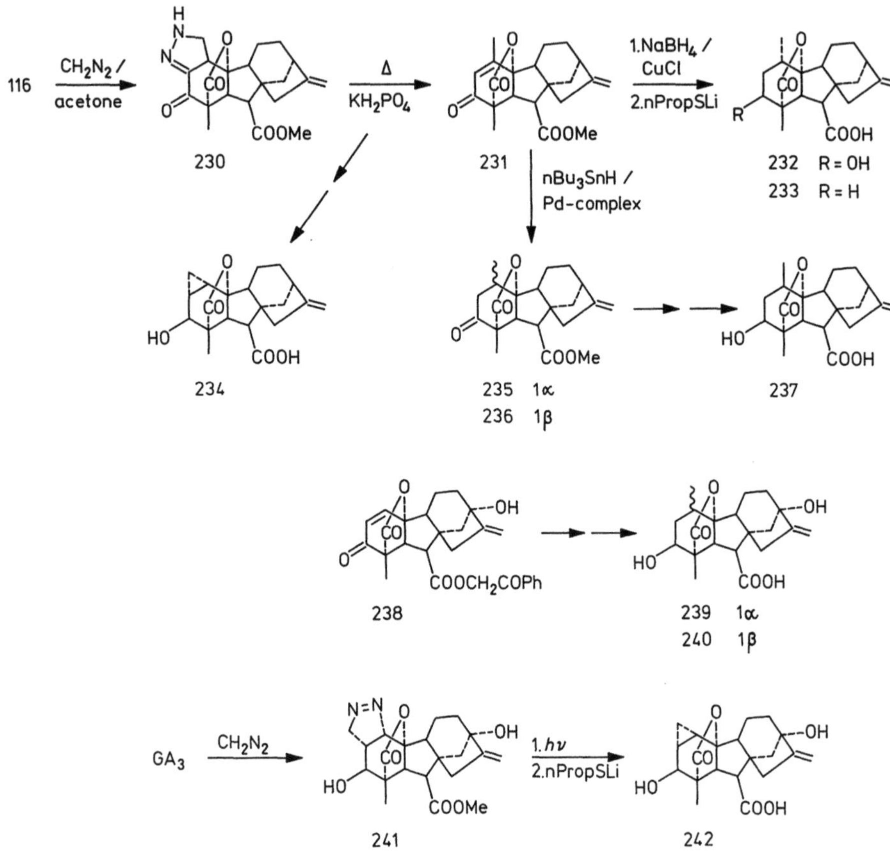

paration and again borohydride reduction pure 1β-methyl GA_4 (**237**) was isolated. From the product of the thermolysis of pyrazoline **230** also a corresponding 1α,2α-cyclopropyl-3-ketone was obtained as a minor product and further transformed in conventional steps to 1α,2α-cyclopropyl GA_4 (**234**). In analogous sequences starting from the 3-dehydro-GA_3 phenacyl ester (**238**) 1α-methyl GA_1 (**239**), 1β-methyl GA_1 (**240**) and 1α,2α-cyclopropyl GA_1 (**242**) were prepared [208].

Voigt et al. published a more efficient route to **242** in only three steps starting from GA_3. Thus, under long-termed conditions direct addition of CH_2N_2 to the allylic Δ^1-double bond of GA_3 took place to afford the 1α,2α-pyrazoline ester **241** which was photolyzed and methylated to **242** in moderate overall-yield [209].

In biological studies of the synthesized gibberellin alkyl analogs 2,2-dimethyl GA_4 (**201**) and 2,2-dimethyl GA_1 (**202**) were found to be 10–100 times more active than the parent hormones representing striking examples for increase of gibberellin bioactivity by structural modification [210].

5.4 Nitrogen- and Sulphur-Containing Gibberellins

Up to now there are only limited systematical studies on N- and S-containing gibberellin compounds. Thus, a wide number of GA_1- and GA_3-7-amides of type **245** and **246** have been prepared via aminolysis of the corresponding gibberellin anhydrides **243** and **244** with aliphatic or aromatic amines [211]. Reaction with diamines led to basic amides **247** and **248**, respectively. For example, with putrescine (n = 4) the corresponding 4-amino-*n*-butylamides were obtained and used as suitable intermediates for gibberellin affinity columns [125, 212] and radioimmunoassay [213].

Similar, Adam et al. reported the synthesis of nitrogen-containing GA_1 and GA_3 conjugates of type **251** and **252**, respectively. This is either by direct fission of **243** and **244** with alkali salts of L-amino acids or by aminolysis with esters of L-amino acids to give **249** and **250** and then by Bartlett fission to give **251** and **252**. In such a manner, GA_1-oylglycine, -glycylglycine and -proline as well as GA_3-oylglycine, -glycylglycine, -alanine, -valine, -leucine, -phenylalanine, -serine, -proline, -tyrosine and aspartic acid were prepared [214]. Although there are hints concerning a native occurrence of such amphoteric gibberellin amino acid conjugates [215] so far no defined member has been isolated from plants.

Upon reaction of 3-dehydro-GA_1 (**253**) with amines under azeotropic conditions via a retro-Claisen mechanism smoothly 3,4-*seco*-3-amides of type **254** were formed [216]. Under same conditions gibberellin C (**255**), available by Wagner-Meerwein rearrangement of GA_1, underwent condensation to corresponding azomethines which can be reduced stereoselectively with borohydride to give amino gibberellins of structure **256** [217]. 1β-Amino-GA_1 (**257**) is available by

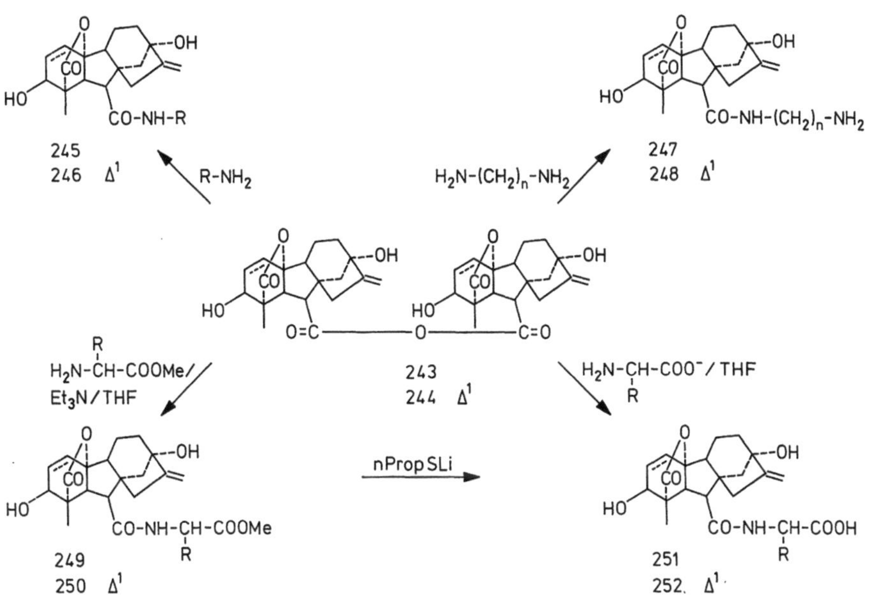

reduction of 1β-azido GA_1 (**102**, see Sect. 3.22) with propane-1,3-dithiol/triethylamine [218].

For preparation of compounds suitable for studying the design of molecular probes the methyl ester of 3-dehydro-GA_3 (**13**) was reacted with bifunctionalized mercaptanes to afford under addition to the Δ^1 double bond preferentially the 1β-mercapto gibberellins **258–262**. The reaction enables the introduction of 1-substituents of varying chain length and different terminal function suitable for further transformations directed to photoaffinity label, gibberellin antigens and affinity columns [219]. Later it has been found that solid-phase acid-catalysis provides stereoselectively the corresponding 1α-epimers. Other mercapto derivatives suitable for molecular probes are compounds of structures **263** and **264** prepared by free radical addition of dithiols to the terminal methylene group of GA_4 and GA_9 [172].

So far two types of true sulphur-containing gibberellin phytohormone analogs have been synthesized by Schierhorn et al. for biological studies [220]. For preparation of 7-thiolo-gibberellins again the anhydrides **243** and **244** were useful intermediates which afforded upon sulfhydrolysis 7-thiolo-GA_1 (**265**) and 7-thiolo-GA_3 (**266**), respectively. Photolysis of this thiogibberellins led under intramolecular functionalization to the corresponding 7,15α-thiolactones **267** and **268**. Compounds **265** and **266** were found to be instable and underwent hydrolysis even under bioassay conditions. For synthesis of the stable 19,10-thiolo-gibberellin **272** the following reaction sequence has been developed: 3,13-diacetyl-$\Delta^{1(10)}$-19-oic acid **72** was transformed via its chloride **269** and sulphydrolysis to the 19-thiolo acid **270**. Upon UV-irradiation of **270** in THF regio- and stereoselective intra-

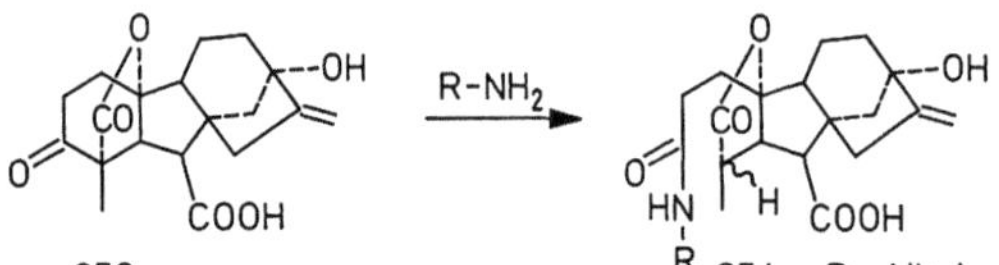

253   254 R = Alkyl

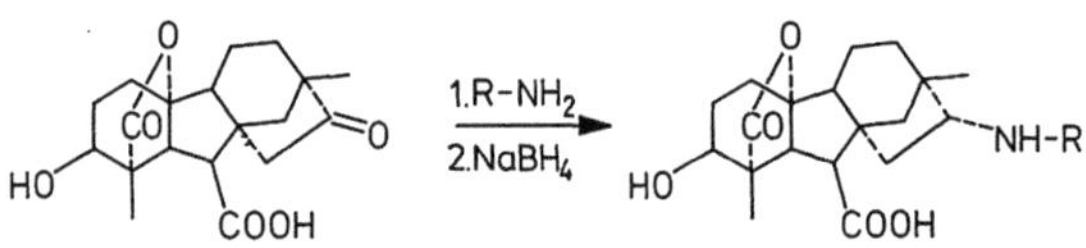

255   256 R = Alkyl

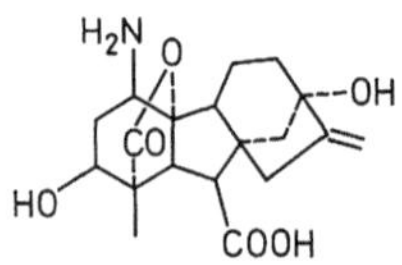

257

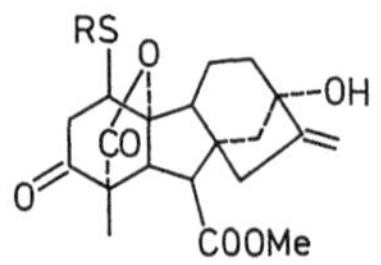

258 R = Alkyl
259 R = -$(CH_2)_2$-OH
260 R = -$(CH_2)_2$-NH_2
261 R = -$(CH_2)_2$-SH
262 R = -CH_2-COOH

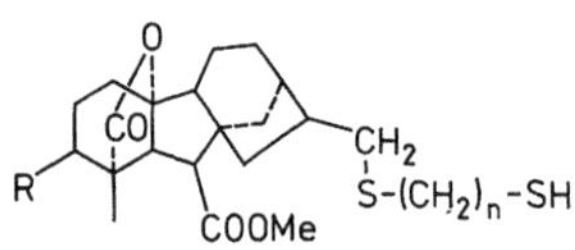

263 R = OH
264 R = H

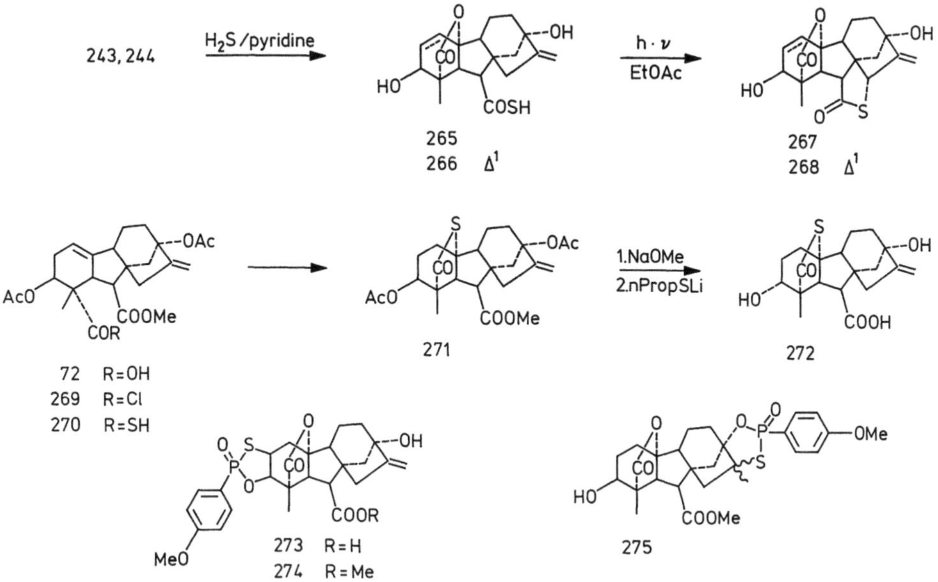

molecular photoaddition of the thiocarboxy function to the $\Delta^{1(10)}$ double bond took place leading to the desired γ-thiolactone **271**. Compound **270** could also be cyclized to **271** upon treatment with H_2S/pyridine. Alkaline hydrolysis of **271** under simultaneous inversion at C-3 via a retro-aldol mechanism followed by demethylation gave 19,10-thiolo-3-*epi*-GA_1 (**272**). In the dwarf maize bioassay this thioanalog showed an activity higher than found for GA_1. Experiments towards direct thiation of GA_3, GA_3 methyl ester (**6**) or GA_1 methyl ester (**7**) with Lawesson reagent led to the $2\beta,3\beta$- and 13,16-annulated oxathiaphospholanes **273**, **274** and **275**, respectively [221].

The only one hitherto known native sulphur-containing gibberellin is gibberethione (**276**), a catabolic product of GA_3 isolated from seed of *Pharbitis nil* [222]. Its synthesis was achieved by reaction of 3-dehydro-GA_3 [**13**] with ammonium mercaptopyruvate at pH 8 to afford in a direct coupling the desired **276** and *iso*-gibberethione **277** in a 2:5,4 ratio. The analogous reaction starting from 3-dehydro-GA_7 led to the corresponding 13-deoxy derivatives of **276** and **277** [223].

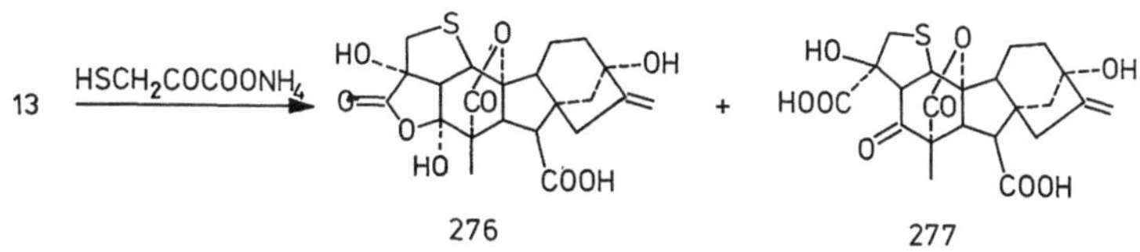

6 Conclusions

The object of gibberellins as biologically active endogenous hormones in the higher plants has stimulated considerable chemical research in this field. Improved analytical techniques encouraged the discovery of new native gibberellins which has led to the identification of 74 members from higher plants but also from fungi. Synthetic work on this group of plant hormones is growing quite rapidly. Whereas total synthesis of gibberellins till now has had more theoretical interest, gibberellin A_3, produced on a large scale by fermentation with *Gibberella fujikuroi*, is a cheap and suitable starting material for manifold efforts in partial synthesis. One main line of this research is directed to the preparation of scarce native gibberellins needed badly for further biological studies. These activities also include the synthesis of labelled gibberellins needed for investigation of biosynthesis, localization, transport, metabolism, and studies on the mode of action. A further main direction in chemical gibberellin research is represented by synthesis of new structure analogs with the aim of getting a deeper insight into structure-activity relationships and mode of action, as well as to use such compounds as biochemical models, molecular probes and as potential plant growth regulators. The recent developments shown in this review indicate that the applied synthetic concepts are being rationalized more and more by biochemical reflexions. As a result of these efforts new synthetic strategies to biologically active gibberellins have been developed leading to first analogs with activity higher than the parent hormones as well as to compounds which act as gibberellin inhibitors. Further intensive interdisciplinary research involving chemical, biochemical, and biological studies is necessary to understand better the phytohormonal function of the gibberellins and to broaden the scope of practical application in agriculture and horticulture.

Acknowledgement. The authors are indebted to Mr. E. Schäfer, Mrs. G. Schmidt and Mrs. M. Süsse for accurate drawing of the formulas and Mrs. H. Bottin for typing the manuscript.

7 References

(Recent literature added in proof is given in refs. [224]–[235])

1. Graebe JE, Ropers HJ (1978) Gibberellins. In: Letham DS, Goodwin PB, Higgins TJV (eds) Phytohormones and related compounds — A Comprehensive Treatise, vol. I. Elsevier, Amsterdam, p 107
2. Crozier A (ed) (1983) The biochemistry and physiology of gibberellins, vols 1 and 2, Praeger, New York
3. Rowe JW (1968) The common and systematic nomenclature of cyclic diterpenes, 3rd Revision, US Forest Product Laboratory, Madison, WI
4. MacMillan J, Takahashi N (1968) Nature (London) 217:170

5. Bearder JR (1980) Plant hormones and other growth substances — their background, structures and occurrence. In: MacMillan J (ed) Hormonal regulation of development. Springer, Berlin Heidelberg New York, p 9 (Encyclopedia of Plant Physiology vol 9)
6. Muromtsev G, Agnistikova VN (1984) Gibberellins. Nauka, Moscow
7. Takahashi N, Yamaguchi I, Yamane H (1986) Gibberellins. In: Takahashi N (ed) Chemistry of plant hormones. CRC, Boca Raton, FL, p 57
8. Schneider G (1987) Gibberellin conjugation. In: Schreiber K, Schütte HR, Sembdner G (eds) Conjugated plant hormones — structure, metabolism, and function. Institute of Plant Biochemistry Halle, Academy of Sciences GDR p 158 (Proceedings of the International Symposium Conjugated Plant Hormones, 3–7 Nov 1986 at Gera)
9. Sembdner G, Schneider G, Schreiber K (eds) (1987) Methoden der Pflanzenhormonanalyse. Springer, Berlin Heidelberg New York
10. Preiss A, Adam G, Šaman D, Buděšinsky M (1987) Magn. Reson. Chem. 25: 239
11. Takahashi N, Kitamura H, Kawarada A, Seta Y, Takai M, Tamura S, Sumiki Y (1955) Bull. Agr. Chem. Soc. Japan 19: 267
12. Grove JF, Jeffs PW, Mulholland TPC (1958) J. Chem. Soc. 1236
13. MacMillan J, Seaton JC, Suter PJ (1960) Tetrahedron 11: 60
14. Grove JF (1961) J. Chem. Soc. 3545
15. Curtis PJ, Cross BE (1954) Chem. Ind. (London) 1066
16. Cross BE, Grove JF, MacMillan J, Moffatt J, Mulholland TPC, Seaton JC, Sheppard N (1959) Proc. Chem. Soc. (London) 302
17. Takahashi N, Seta Y, Kitamura H, Sumiki Y (1957) Bull. Agr. Chem. Soc. Japan 21: 396
18. Takahashi N, Seta Y, Kitamura H, Sumiki Y (1959) Bull. Agric. Chem. Soc. Japan 23: 405
19. West CA, Phinney BO (1959) J. Am. Chem. Soc. 81: 2424
20. MacMillan J, Pryce RJ (1967) J. Chem. Soc. 550
21. MacMillan J, Seaton JC (1962) Tetrahedron 18: 349
22. Cross BE, Galt RHB, Hanson JR (1960) Tetrahedron Lett. 15: 18
23. Cross BE, Galt RHB, Hanson JR (1962) Tetrahedron 18: 451
24. Hanson JR (1966) Tetrahedron 22: 701
25. Brown JC, Cross BE, Hanson JR (1967) Tetrahedron 23: 4095
26. Cross BE, Norton HJ (1965) J. Chem. Soc. 1570
27. Galt RHB (1965) J. Chem. Soc. 3143
28. Cross BE (1966) J. Chem. Soc. 501
29. Hanson (1967) Tetrahedron 23: 733
30. Galt RHB (1968) Tetrahedron 24: 1337
31. Bearder JR, MacMillan J (1973) J. Chem. Soc. Perkin Trans 1 22: 2824
32. Pryce RJ, MacMillan J (1967) Tetrahedron Lett. 42: 4173
33. Koshimizu K, Fukui M, Kusaki M (1966) Tetrahedron Lett. 22: 2459
34. Koshimizu K, Ishii H, Fukui M, Mitsui T (1972) Phytochemistry 11: 2355
35. Murofushi N, Iriuchijima S, Takahashi N (1966) Agric. Biol. Chem. 30: 917
36. Murofushi N, Takahashi N, Yokota T, Tamura S (1968) Agric. Biol. Chem. 32: 1239
37. Takahashi N, Murofushi N, Yokota T (1967) Tetrahedron Lett. 48: 4861
38. Koshimizu K, Ishii H, Ogawa Y, Mitsui T (1968) Tetrahedron Lett. 9: 1143
39. Harrison DM, MacMillan J, Galt RHB (1968) Tetrahedron Lett. 27: 3137
40. Harrison DM, MacMillan J (1971) J. Chem. Soc. 631
41. Yokota T, Takahashi N, Murofushi N, Tamura S (1969) Planta 87: 180
42. Takahashi N, Yokota T, Murofushi N, Tamura S (1969) Tetrahedron Lett. 25: 2077
43. Fukui H, Koshimizu K, Mitsui T (1971) Phytochemistry 10: 671
44. Yokota T, Murofushi N, Takahashi N, Tamura S (1971) Agric. Biol. Chem. 35: 583
45. Murofushi N, Yokota T, Takahashi N (1970) Agric. Biol. Chem. 34: 1436
46. Yamaguchi Y, Yokota T, Murofushi N, Takahashi N (1975) Agric. Biol. Chem. 39: 2405
47. Murofushi N, Yokota T, Takahashi N (1971) Agric. Biol. Chem. 35: 441
48. Yamane H, Yamaguchi I, Murofushi N, Takahashi N (1971) Agric. Biol. Chem. 35: 1144
49. Bearder JR, MacMillan J (1972) Agric. Biol. Chem. 36: 342
50. Bearder JR, MacMillan J (1973) J. Chem Soc. Perkin Trans 1 2824
51. Hiraga K, Yokota T, Murofushi N, Takahashi N (1972) Agric. Biol. Chem. 36: 345

52. Fukui H, Nemori R, Koshimizu K, Yamazaki Y (1977) Agric. Biol. Chem. 41: 175
53. Yamane H, Fujioka S, Spray CR, Phinney BO, MacMillan J, Gaskin P, Takahashi N (1988) Plant Physiol. 86: 857
54. Beeley LJ, Gaskin P, MacMillan J (1975) Phytochemistry 14: 779
55. Frydman VM, Gaskin P, MacMillan J (1974) Planta 118: 123
56. Bearder JR, Dennis FG, MacMillan J (1975) Tetrahedron Lett. 9: 669
57. Beeley LJ, MacMillan J (1976) J. Chem. Soc. Perkin Trans. 1 1022
58. Fukui H, Koshimizu K, Nemori R (1978) Agric. Biol. Chem. 42: 1571
59. Sponsel VM, MacMillan J (1977) Planta 135: 129
60. Sponsel VM, Gaskin P, MacMillan J (1979) Planta 146: 101
61. Murofushi N, Sugimoto M, Itoh K, Takahashi N (1979) Agric. Biol. Chem. 43: 2179
62. Gaskin P, Kirkwood PS, Lenton JR, MacMillan J, Radley ME (1980) Agric. Biol. Chem. 44: 1589
63. Murofushi N, Sugimoto M, Itoh K, Takahashi N (1980) Agric. Biol. Chem. 44: 1583
64. Beale MH, Bearder JR, Hedden P, Graebe JE, MacMillan J (1984) Phytochemistry 23: 565
65. Yokota T, Takahashi N (1981) Agric. Biol. Chem. 45: 1251
66. Kirkwood PS, MacMillan J (1982) J. Chem. Soc. Perkin Trans 1 689
67. Dolan SC, Holdup DW, Hutchison M, MacMillan J (1985) J. Chem. Soc. Perkin Trans 1 651
68. Hutchison M, (1983) Ph. D. Thesis Univ. Bristol
69. Dolan SC, MacMillan J (1985) J. Chem. Soc. Perkin Trans. 1 2741
70. Yamane H, Yamaguchi I, Kobayashi M, Takahashi M, Sato Y, Takahashi N, Iwatsuki K, Phinney BO, Spray CR, Gaskin P, MacMillan J (1985) Plant Physiol. 78: 899
71. MacMillan J (1985) Hormones: Synthesis, metabolism and biochemistry. In: Randall DD (ed) Current topics in plant biochem. and physiol., Univ. of Missouri, Columbia Press, MO, vol 4 p 53
72. Yamane H, Sato Y, Nohara K, Nakayama M, Murofushi N, Takahashi N, Takeno K, Furuya M, Furber M, Mander LN (1988) Tetrahedron Lett 29: 3959
73. Murofushi N, Nakayama M, Takahashi N, Gaskin P, MacMillan J (1988) Agric. Biol. Chem. 52: 1825
74. Mander LN (1988) Nat. Prod. Reports 541
75. Adam G, Bergner C, Lischewski M, Schierhorn A, Sembdner G, Voigt B (1986) Gibberellins — Synthesis and bioactivity In: Harms AF (ed) Innovative approaches in drug research. Elsevier, Amsterdam (Proceedings of the 3rd Noordwijkerhout Symposium on Medicinal Chemistry, 3–6 September 1985) p 171
76. Kutschabsky L, Adam G (1983) J. Chem. Soc. Perkin Trans. 1 1653
77. Moffatt JS (1960) J. Chem. Soc. 3045
78. Mulholland TPC (1958) J. Chem. Soc. 2693
79. Pryce RP (1973) Phytochemistry 12: 507
80. Cross BE, Grove JF, MacMillan J, Mulholland TPC (1958) J. Chem. Soc. 2520
81. Cross BE, Grove JF, Morrison A (1961) J. Chem. Soc. 2498
82. Kirkwood PS, MacMillan J, Sinnott ML (1980) J. Chem. Soc. Perkin Trans. 1 2117
83. Hsü Y, Takahashi N, Miya K, Kawarada A, Kitamura H, Tamura S, Sumiki Y (1961) Agric. Biol. Chem. 25: 865
84. Jones DF, McCloskey P (1963) J. Appl. Chem. 13: 324
85. Hung PD (1974) Ph. D. Thesis Univ. Halle, GDR
86. Murofushi N, Yamaguchi I, Ishigooka H, Takahashi N (1976) Agric. Biol. Chem. 40: 2471
87. Corey EJ, Brennan TM, Carney RL (1971) J. Am. Chem. Soc. 93: 7316
88. Bartlett PA, Johnson WS (1970) Tetrahedron Lett. 51 :4459
89. Yamaguchi I, Takahashi N, Fujita K (1975) J. Chem. Soc. Perkin Trans. 1 992
90. Minh NV (1979) Ph. D. Thesis Univ. Halle, GDR
91. MacMillan J, Willis CL (1986) J. Chem. Soc. Perkin Trans. 1 309
92. Murofushi N, Durley RC, Pharis RP (1977) Agric. Biol. Chem. 41: 1075
93. Schreiber K, Schneider G, Sembdner G (1966) Tetrahedron 22: 1437
94. Avent AG, Baynham MK, Hanson JR, Hitchcock PB, de Oliveira BH (1989) J. Chem. Soc. Perkin Trans. 1 627

95. Kirkwood JS, MacMillan J, Beale MH (1982) J. Chem. Soc. Perkin Trans. 1 699
96. Lombardo L (1982) Tetrahedron Lett. 23: 4293
97. Cross BE (1960) J. Chem. Soc. 2044
98. Kobrina NS, Serebryakov EP, Kucherov VF, Adam G, Voigt B (1973) Tetrahedron 29: 3425
99. MacMillan J, Willis CL (1984) J. Chem. Soc. Perkin Trans. 1 357
100. Duri ZJ, Fraga BM, Hanson JR (1981) J. Chem. Soc. Perkin Trans. 1 161
101. Voigt B, Adam G, Kobrina NS, Serebryakov EP (1977) Z. Chem. 10: 372
102. Bell RA, Turner JV (1981) Tetrahedron Lett. 22: 4871
103. Brian PW, Grove JF, MacMillan J (1960) Fortschr. Chem. Org. Naturst. 18: 350
104. Abonamer KM, Hanson JR, Hitchcock PB (1987) J. Chem. Soc. Perkin Trans. 1 1991
105. Dolan SC, MacMillan J (1985) J. Chem. Soc. Chem. Commun. 1588
106. Voigt B, Adam G, Häckert H, Jasche K (1988) Ger. (East) patent applied for
107. Bearder JR, MacMillan J (1976) J. Chem. Soc. Chem. Commun. 421
108. Lombardo L, Mander LN, Turner JV (1981) Aust. J. Chem. 34: 745
109. Beale MH, Gaskin P, Kirkwood PS, MacMillan J (1980) J. Chem. Soc. Perkin Trans 1 885
110. Bearder JR, Kirkwood PS, MacMillan J (1981) J. Chem. Soc. Perkin Trans. 1 672
111. Duri ZJ, Fraga BM, Hanson JR (1981) J. Chem. Soc. Perkin Trans. 1 3016
112. Duri ZJ, Hanson JR (1984) J. Chem. Soc. Perkin Trans. 1 603
113. Beale MH, MacMillan J (1979) European patent 0007240
114. Dawe RD, Mander LN, Turner JV (1985) Tetrahedron Lett. 26: 363
115. Dawe RD, Mander LN, Turner JV, Xinfu P (1985) Tetrahedron Lett. 26: 5725
116. Chu A, Mander LN (1988) Tetrahedron Lett. 29: 2727
117. Willis CL (1987) Tetrahedron Lett. 28: 6705
118. Adam G, Hung PD (1974) Tetrahedron Lett. 38: 3419
119. Voigt B, Adam G (1983) Tetrahedron 39: 449
120. Cross BE (1968) In: Reinhold L and Liwschitz Y (eds) Progress in phytochemistry, vol 1, Wiley Interscience, New York p 195
121. Hanson JR (1971) Progr. Chem. Org. Natur. Comp. 29: 395
122. MacMillan J (1971) Diterpenes-the gibberellins In: Goodwin TW (ed) Aspects of terpenoid chemistry and biochemistry. Academic Press, New York, p 153
123. MacMillan J (1974) The Chemistry and Biochemistry of Plant Hormones. In: Runeckles VC, Sondheimer E, Walton DC (eds) Recent Advances in Phytochemistry vol 7, Academic, New York p 1
124. West CA (1973) Biosynthesis of gibberellins. In: Milborrow (ed) Biosynthesis and its control in plants. Academic Press, New York, p 143
125. Knöfel HD, Müller P, Kramell R, Sembdner G (1975) FEBS Letters 60: 39
126. Hanson JR (1980) The partial synthesis and labelling of gibberellins. In: Lenton (ed) Gibberellins — Chemistry, physiology and use. British Plant Regulator Group-Monograph 5, ARC Letcombe Laboratory, Wantage, p 5
127. MacMillan (1979) Partial synthesis of isotopically labelled gibberellins. In: Shoog F (ed) Proc. 10th Int. Conf. on Plant Growth Subst., Madison WO, Springer, Berlin Heidelberg New York, p 161
128. Takahashi N, Yamaguchi I (1983) Chemical procedures for isotopic labelling of gibberellins and ent-kaurane derivatives. In: Crozier A (ed) The biochemistry and physiology of gibberellins, vol. 1. Praeger, New York, p 457
129. Cross BE, Galt RHB, Hanson JR (1964) J. Chem. Soc. 295
130. Birch AJ, Richards RW, Smith H, Harris A, Whalley WB (1959) Tetrahedron 7: 241
131. Pitel DW, Vining LC, Arsenault GP (1971) Can. J. Biochem. 49: 185
132. Sembdner G (1969) Ph. D. Thesis Univ. Halle, GDR
133. Hanson JR, Hawker J (1973) Phytochemistry 12: 1073
134. Verbiscar AJ, Cragg G, Geissmann TA, Phinney BC (1967) Phytochemistry 6: 807
135. Cross BE, Norton K, Stewart JC (1968) J. Chem. Soc. 1054
136. Lin JT, Heftmann E (1979) J. Lab. Comp. 16: 537
137. Kende H (1967) Plant Physiol. 42: 1612
138. Pitel DW, Vining LC (1970) Canad. J. Biochem. 48: 259
139. Hiraga K, Takahashi N (1974) Phytochemistry 13: 2371

140. Phillips IDJ (1974) Planta 116: 109
141. Nadeau R, Rappaport L (1974) Phytochemistry 13: 1537
142. Durley RC, Pharis RP (1973) Planta 109: 357
143. MacMillan, Willis CL (1984) J. Chem. Soc. 351
144. Yokota T, Reeve DR, Crozier A (1976) Agric. Biol. Chem. 40: 2091
145. Raiton ID (1976) J. South. Afr. Bot. 42: 147
146. Murofushi N, Durley RC, Pharis RP (1974) Agric. Biol. Chem. 38: 475
147. Albone KS, MacMillan J, Pitt AR, Willis CL (1986) Tetrahedron 42: 3203
148. Willis CL, Gaskin P, MacMillan J (1988) Phytochemistry 27: 3970
149. Musgrave A, Kende H (1970) Plant Physiol. 45: 56
150. Durley RC, Railton ID, Pharis RP (1973) Phytochemistry 12: 1609
151. Wilzbach KJ (1957) J. Amer. Chem. Soc. 79: 1013
152. Rosenblum C (1959) Nucleonics 17: 80
153. Baumgartner WE, Lazer LS, Dalziel AM, Cardinal EV, Varner EL (1959) Agric. Food Chem. 7:422
154. Lischewski M, Adam G, Liebisch HW, Pleiß U (1982) J. Lab. Comp. 19: 725
155. Lischewski M, Pleiß U (1983) J. Lab. Comp. 20: 1413
156. Lischewski M, Adam G (1980) 36: 1237
157. Lischewski M, Adam G (1978) Ger. DE 2911618
158. Lischewski M (1984) Z. Chem. 24: 64
159. Lischewski M, Adam G, Serebryakov EP (1978) Ger. (East) DD 148 881
160. Lischewski M (1984) Z. Chem. 24: 152
161. Bearder JR, MacMillan J, Phinney BO (1973) Phytochemistry 12: 2173
162. Hanson JR, Parry KP, Willis CL (1982) Phytochemistry 21: 1575
163. Hedden P, MacMillan J, Phinney BO (1974) J. Chem. Soc. 587
164. Reid DM, Tuing MS, Durley RC, Railton ID (1972) Planta 108: 67
165. Bearder JR, Frydman VM, Gaskin P, Hatton IK, Harvey WE, MacMillan J, Phinney BO (1976) J. Chem. Soc. 178
166. Durley RC, Railton ID, Pharis RP (1974) Phytochemistry 13: 547
167. Ingram TJ, Reid JP, Murfet IC, Gaskin P, Willis CL, MacMillan J (1984) Planta 160: 455
168. Duri ZJ, Hanson JR (1984) J. Chem. Soc. 363
169. Beale MH, MacMillan J (1981) J. Chem. Soc 394
170. Down GJ, Lee M, MacMillan J, Staples KS (1983) J. Chem. Soc. 1103
171. Schneider G, Sembdner G, Schreiber K (1965) Gibberelline, ihre Derivate und Abbauprodukte — Eine tabellarische Übersicht. In: Danert S, Metzner P, Müller-Stoll WR, Rieth A, Sagromsky H, Schreiber K, Stubbe H (eds) Die Kulturpflanze vol XIII, Akademie-Verlag, Berlin, p 267
172. Beale MH, Hooley R, MacMillan J (1986) Gibberellins: Structure activity relationships and the design of molecular probes. In: Bopp N (ed) Plant growth substances 1985. Springer, Berlin Heidelberg New York, p 65
173. Serebryakov EP, Agnistikova VN, Suslova LM (1984) Phytochemistry 23: 1847
174. Srivastava LM (1987) The gibberellin receptor. In: Klämbt D (ed) Plant hormone receptors. Proceedings of a NATO advanced research workshop, 18–22 Aug. 1986, Bad Honnef, Springer, Berlin Heidelberg New York, p 199
175. Adam G, Hung PD, Kutschabsky L, Lischewski M, Preiss A, Sung TV (1987) Photochemical reactions as a tool in natural products synthesis — applications in the gibberellin field. In: Chizov O (ed)
Organic synthesis — modern trends. Proceedings of the 6th IUPA symposium on organic synthesis, 10–13 Aug. 1986 Moscow, Blackwell, Oxford, p 65
176. Stoddart JL, Venis MA (1980) Molecular and subcellular aspects of hormone action. In: MacMillan J (ed)
Hormonal regulation of development. Springer, Berlin Heidelberg New York, p 445 (Encyclopädia of Plant Physiology vol 9)
177. Lischewski M, Adam G (1980) Tetrahedron Lett. 21: 1627
178. Serebryakov EP, Lischewski M, Adam G (1978) Izv. Akad. Nauk SSSR, Ser. Chim. 9: 2181
179. Lischewski M (1982) Z. Chem. 22: 141

180. Adam G, Gabb B, unpublished
181. Bergner C, Lischewski M, Adam G, Sembdner G (1982) Planta 155: 231
182. Hanson JR, Willis CL, Parry KP (1984) J. Chem. Soc. Perkin Trans. 1 3020
183. Hanson JR, Parry KP, Triana J, Willis CL (1982) J. Chem. Soc. Perkin Trans. 1 192
184. Bearder JR, MacMillan J, Phinney BO, Hanson JR, Rivett DEA, Willis CL (1982) Phytochemistry 21: 2225
185. Fraga BM, Hernandez MG, Tellado FG (1986) J. Chem. Soc. Perkin Trans. 1 21
186. Bateson JH, Cross BE (1972) J. Chem. Soc. Chem. Commun. 649
187. Bateson JH, Cross BE (1974) J. Chem. Soc. Perkin Trans 1 1131
188. Stoddart JL (1972) Planta 107: 81
189. Cross BE, Erasmuson A (1981) J. Chem. Soc. Perkin Trans. 1 1918
190. Cross BE, Filippone P (1981) J. Chem. Res. (S) 166
191. Bateson JH, Cross BE (1974) J. Chem. Soc. Perkin Trans. 1 2409
192. Banks RE, Cross BE (1977) J. Chem. Soc. Perkin Trans, 1 512
193. Boulton K, Cross BE (1981) J. Chem. Soc. Perkin Trans. 1 427
194. Jones TWA (1976) Phytochemistry 15: 1825
195. Beale MH, MacMillan J (1981) Phytochemistry 20: 693
196. Hoad GV, Phinney BO, Sponsel VM, MacMillan J (1981) Phytochemistry 20: 703
197. Cross BE, Simpson IC (1981) J. Chem. Soc. Perkin Trans. 1 98
198. Keith B, Srivastava LM, Murofushi N (1979) Agric. Biol. Chem. 43: 141
199. Voigt B, Adam G (1988) Z. Chem. 28: 60
200. Adam G, Sung TV Ger. (East) DD 134 643
201. Hung PD, Adam G Ger. (East) DD 229 127
202. Hoad GV, MacMillan J, Smith VA, Sponsel VM, Taylor DA (1982) Gibberellin 2β-hydroxylases and Biological Activity of 2β-Alkyl Gibberellins. In: Wareing PE (ed) Plant growth substances. Academic, London, p 91
203. Beale MH, MacMillan J, Spray CR, Taylor DA (1984) J. Chem. Soc. Perkin Trans. 1 541
204. MacMillan J, Taylor DA (1985) J. Chem. Soc. Perkin Trans. 1 837
205. Beale MH (1985) J. Chem. Soc. Perkin Trans. 1 1151
206. Fowles AM, Beale MH, Jones DNM, MacMillan J, Willis C (1988) J. Chem. Soc. Perkin Trans. 1 1983
207. Fowles AM, MacMillan J (1988) J. Chem. Soc. Perkin Trans. 1 1973
208. MacMillan J, Willis C (1985) J. Chem. Soc. Perkin Trans. 1 2177
209. Voigt B, Adam G, Kutschabsky L, Ger. (East) DD 237 654
210. Hoad GV, Phinney BO, Sponsel VM, MacMillan J (191) Phytochemistry 20: 703
211. Adam G, Lischewski M, Sych FJ, Ulrich A (1976) J. Prakt. Chem. 318: 105
212. Müller D, Knöfel HD, Kramell R (1976) Z. Chem. 16: 105
213. Weiler EW, Wieczorek U (1984) Planta 152: 159
214. Adam G, Lischewski M, Sych FJ, Ulrich A (1977) Tetrahedron 33: 95
215. Nadeau R, Rappaport L (1974) Plant Physiol. 54: 809
216. Adam G, Sung TV (1987) Z. Chem. 27: 146
217. Hung PD, Adam G (1984) J. Prakt. Chem. 326: 253
218. Voigt B, Adam G (1983) Z. Chem. 23: 177
219. Voigt B, Adam G (1979) Pharmazie 34: 362
220. Schierhorn A, Adam G, Kutschabsky L, Leibnitz P (1987) J. Chem. Soc. Perkin Trans. 1 2111
221. Schierhorn A (1983) Ph. D. Thesis, Academy of Sciences of G.D.R. Halle
222. Yokota T, Yamazaki S, Takahashi N, Iitaka Y (1974) Tetrahedron Lett. 34: 2957
223. Yokota T, Yamane H, Takahashi N (1976) Agric. Biol. Chem. 40: 2507

Recent Literature Added in Proof

Structure and Occurrence

224. Castellaro SJ, MacMillan J, Singh AK, Willis CL (1990) J. Chem. Soc. Perkin Trans. 1 145
225. Lenton JR, Appleford NEJ (1990) Gibberellin production and action during germination of wheat. In: Takahashi N, Pinney BO, MacMillan (eds) Proceedings of Gibberellin Symposium Tokyo 1989, Springer 1990, in press

Synthesis

226. Avent AG, Baynham MK, Hanson JR, Hitchcock PB, de Oliveira BH (1989) J. Chem. Soc. Perkin Trans. 1 627
227. Sassa T, Suzuki K, Haruki E (1989) Agric. Biol. Chem. 53: 303
228. Nagaoka H, Shimano M, Yamada Y (1989) Tetrahedron Lett. 30: 971
229. Voigt B, Porzel A, Adam G, Kutschabsky L (1989) Z. Chem. 29: 174
230. Fraga BM, Hanson JR, Hernandez MG, Tellado FG (1989) Tetrahedron Lett. 30: 6899
231. Mander LN, Patrick GL (1990) Tetrahedron Lett. 31: 423
232. Wang ZM, Zhou WS (1990) Chinese J. Chem. 1: 84
233. Nester-Hudson JE, Semenenko FM, Beale MH, MacMillan J (1990) Phytochemistry 29: 1041
234. Lischewski M, Knoefel H-D (1990) Z. Chem. 30: 58
235. Shimano M, Nagaoka H, Yamada Y (1990) Chem. Pharm. Bull. 38: 276
236. Nakayama M, Yamane H, Yokota T, Yamaguchi I, Murofushi N, Takahashi N, Nishijima T, Katsura N, Nonaka M, Gaskin P, MacMillan J, Mander LN, Chu A (1990) Agric. Biol. Chem. 54: 837
237. Watanabe M, Sakai Y, Teraoka T, Abe H, Kono Y, Uzawa J, Kobayashi K, Suzuki Y, Sakurai A (1990) Agric. Biol. Chem. 54: 1103
238. Fraga BM, Gomez CD, Hernandez MC, dePaz E, Tellado FG, Perales AA (1990) J. Chem. Soc. Perkin Trans. 1 1535
239. Castellaro SJ, Dolan SC, MacMillan J, Willis CL (1990) Phytochemistry 29: 1823
240. Castellaro SJ, Dolan SC, Hedden P, Gaskin P, MacMillan J (1990) Phytochemistry 29: 1833
241. Adam G, Voigt B, Lischewski M, Schierhorn A, Preiss A, Kutschabsky L (1990) New aspects in synthesis of gibberellins. In: Atta-ur-Rahman (ed) Studies in Natural Products Chemistry, Springer 1990, in press
242. Lischewski M (1990) Z. Chem. 30: 290
243. Schneider G, Schreiber K, Jensen E, Phinney BO (1990) Liebigs Ann. Chem. 5: 491
244. Cartenn-Lichterfelde C, Davis CA, Hanson JR (1990) J. Chem. Res. (S) 169
245. Nester-Hudson JE, Semenenko FM, Beale MH, MacMillan J (1990) Phytochemistry 29: 1041
246. Beale MH (1990) J. Chem. Soc. Perkin Trans. 1 925

Recent Advances in Brassinosteroid Research

Volker Marquardt and Günter Adam
Institute of Plant Biochemistry, Weinberg, O-4050 Halle/Saale, FRG

The brassinosteroids represent a class of steroidal plant growth regulators with high growth-promoting activity. Evidence is accumulating that they are of ubiquitous occurrence and constitute a new group of phytohormones. The article gives a comprehensive survey on the hitherto known brassinosteroids isolated from plants, their analysis as well as the synthesis of members of the class and structural analogs, including labelled compounds. Recent advances in the studies on bioactivity, structural-activity relationships and physiology are also discussed. Experiments for practical use indicate a broad spectrum of bioregulatory and antistress effects which seem to prove the brassinosteroids as ecologically beneficial plant regulators for future application in agriculture and horticulture. An up-to-date overview of the research in this rapidly developing field covering the literature up to 1990 is given.

Chemistry of Plant Protection 7

1 Introduction

In 1979 Grove et al. [1] isolated from rape pollen *(Brassica napus)* a highly active plant growth promotor and elucidated its structure as (22*R*,23*R*,24*S*)-2α,3α,22,23)-tetrahydroxy-24-methyl-B-*homo*-7-oxa-5α-cholestan-6-one [1, see Table 1]. The unique structural features of this compound as well as its dramatic ability to stimulate plant growth gave rise to intense chemical and biological research activities in many laboratories. As a result of these efforts evidence is accumulating that brassinosteroids are a new group of plant hormones of substantial practical interest. Various reviews on synthesis [2, 3, 4], analysis [5, 6] and including biological aspects [7–10] of brassinosteroids have been published. In the following, we discuss recent progress in chemistry, bioactivity and application of this group of steroidal plant growth regulators.

2 Structure

Since the discovery of brassinolide (**1**) 26 other naturally occurring brassinosteroids have been structurally elucidated. Their structures and physical constants are given in Table 1. About 40 [11] further members were suggested on the basis of HPLC and MS data however without finally structural determination up till now.

Table 1. Structure of naturally occurring brassinosteroids

Structure	Compound	Mp in °C	$[\alpha]_D$	References
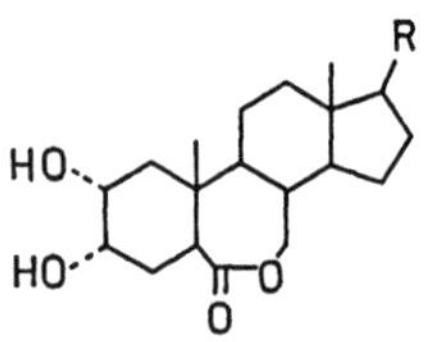				
R = 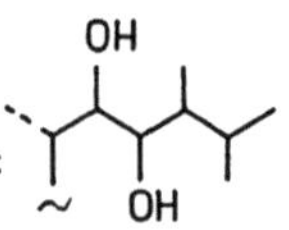	Brassinolide (**1**)	274–275	41.9°	[1, 16]
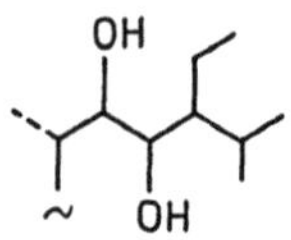	*Homo*brassinolide (**2**)	271–273	33°	[17, 18]

Table 1. (continued)

Structure	Compound	Mp in °C	$[\alpha]_D$	References
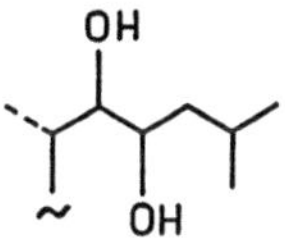	28-*Nor*brassinolide (**3**)	256–259	32°	[17, 19, 20]
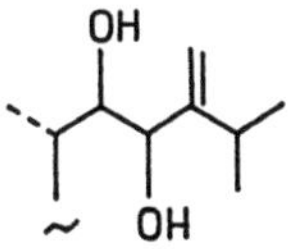	Dolicholide (**4**)	235–237	56.3°	[21, 22, 23]
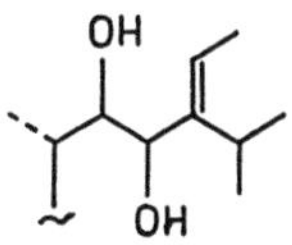	*Homo*dolicholide (**5**)	214–216	35.4°	[22, 24, 25]
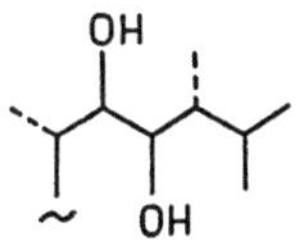	24-*Epi*brassinolide (**6**)	256–258	30°	[26, 27]
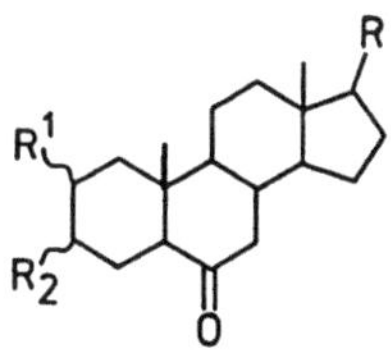				
R = 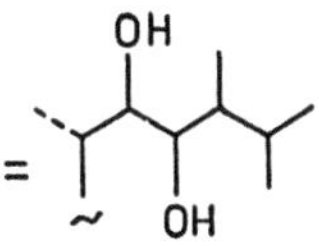	Castasterone (**7**) $R^1 = R^2 = \alpha$-OH	250–261	−4°	[28, 29]
	2-*Epi*castasterone (**8**) $R^1 = \beta$-OH, $R^2 = \alpha$-OH			[15]
	3-*Epi*castasterone (**9**) $R^1 = \alpha$-OH, $R^2 = \beta$-OH			[15]
	2,3-Di*epi*castasterone (**10**) $R^1 = R^2 = \beta$-OH			[15]
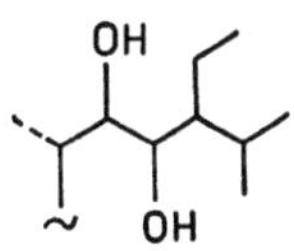	Ethylbrassinone (**11**) $R^1 = R^2 = \alpha$-OH	253–256		[19, 30, 31]

Table 1. (continued)

Structure	Compound	Mp in °C	$[\alpha]_D$	References
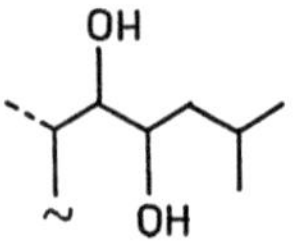	Brassinone (**12**) $R^1 = R^2 = \alpha$-OH	221–222	−14°	[18, 19]
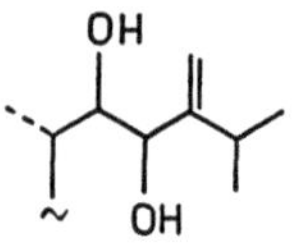	Dolichosterone (**13**) $R^1 = R^2 = \alpha$-OH	230–232.5	4°	[22, 32]
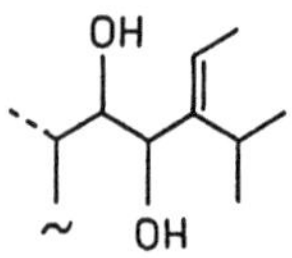	*Homo*dolichosterone (**14**) $R^1 = R^2 = \alpha$-OH	225–226	9.8°	[22, 25]
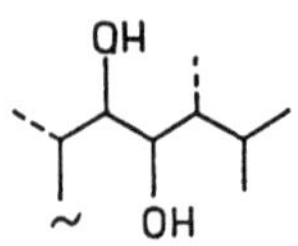	24-*Epi*castasterone (**15**) $R^1 = R^2 = \alpha$-OH	241–242	1°	[30, 33]
	3,24-Di*epi*castasterone (**16**) $R^1 = \alpha$-OH, $R^2 = \beta$-OH			[11]
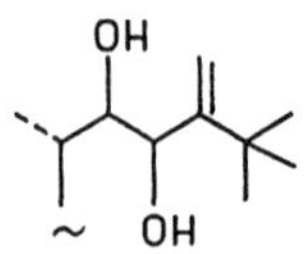	25-Methyldolichosterone (**17**) $R^1 = R^2 = \alpha$-OH	246–247	4.3°	[15, 34]
	2.3-Di*epi*-25-methyl-dolichosterone (**18**) $R^1 = R^2 = \beta$-OH	248	4.6°	[15, 34]
	2-*Epi*-25-methyldolicho-sterone (**19**) $R^1 = \beta$-OH, $R^2 = \alpha$-OH			[11, 15]
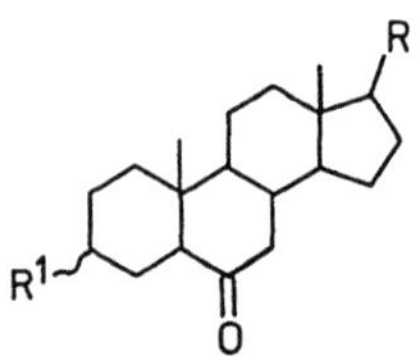				
R = 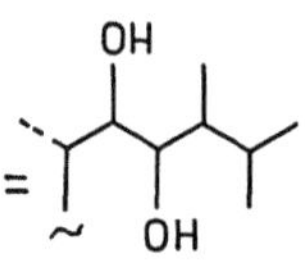	2-Deoxycastasterone (**20**) (Typhasterol) $R^1 = \alpha$-OH	227–230		[35]
	2-Deoxy-3-*epi*castasterone (Teasterone) (**21**) $R^1 = \beta$-OH			[36]

Table 1. (continued)

Structure	Compound	Mp in °C	$[\alpha]_D$	References
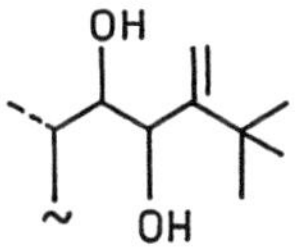	2-Deoxy-25-methyldolichosterone (**22**) $R^1 = \alpha$-OH			[15]
	2-Deoxy-3-*epi*-25-methyldolichosterone (**23**) $R^1 = \beta$-OH			[15]
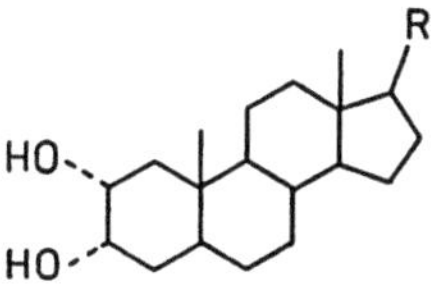				
R = 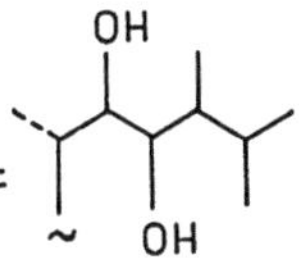	6-Deoxocastasterone (**24**)	225–226		[22, 23, 37, 38]
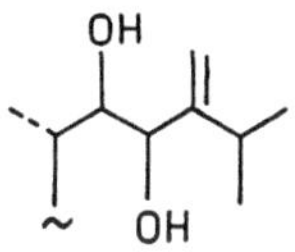	6-Deoxodolichosterone (**25**)	219–220.5	33.2°	[22, 37, 38, 39]
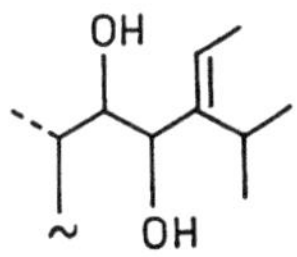	6-Deoxo*homo*dolichosterone (**26**)	225–227		[38, 40]
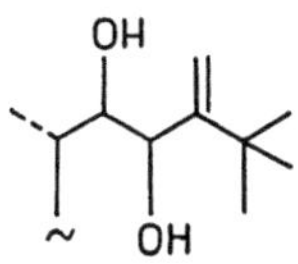	6-Deoxo-25-methyldolichosterone (**27**)			[15]

With regard to the A/B ring functionalities the hitherto clarified members (Table 1) can be divided into the following subgroups:

- Compounds with 7-membered 7-oxalactone-B-ring and vicinal 2α,3α-hydroxy functions (**1**–**6**)

- 6-Oxocompounds with a 6-membered B-ring, having two hydroxy functions in positions 2 and 3 (**7**–**19**)
- 6-Oxo compounds with 3α- or 3β oriented hydroxy function (**20**–**23**) and
- Compounds without oxygen functions in the B-ring (**24**–**27**)

With regard to the side chain moiety all 27 natural brassinosteroids exhibit a characteristic (22*R*,23*R*)-vicinal dihydroxy function. Further structural variations lie in the different alkylation at C-24, reflecting the parent phytosterols which are assumed to be the biogenetic precursors. In the case of the 25-methylated brassinosteroids the corresponding 25-methyl-24-methylenecholesterol is known and has been detected in the same plant source [12, 13].

From *Phaseolus vulgaris* the first brassinosteroid conjugates were isolated and structurally determined as 23-*O*-β-D-glucopyranosyl-25-methyldolichosterone (**28**) [14] and 23-*O*-β-D-glucopyranosyl-2-*epi*-25-methyldolichosterone (**29**) [15].

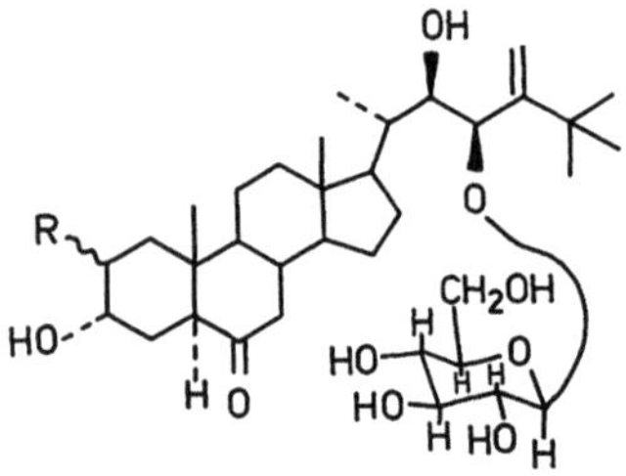

28 R= α-OH
29 R= β-OH

3 Occurrence

The occurrence of native brassinosteroids is summarized in Table 2 (not included brassinosteroids – **8, 9, 10, 16, 18, 19, 22, 23, 27, 29** – which were found in *Phaseolus vulgaris* only), indicating a wide distribution in the plant kingdom. Agriculturally important species such as rape, rice, tea, Chinese cabbage, bean, sunflowers and radish are among the investigated plant sources. Also in other plant families such as Pinaceae, Convolvulaceae, Typhaceae, Betulaceae, Apocynaceae, Fagaceae, Polygonaceae or Hamamelidaceae as well as in the green algae *Hydrodictyon reticulatum* and very recently in a Equisetaceae, brassinosteroids have been detected. Thus, brassinosteroids seem to be of ubiquitous occurrence in plants. They can be isolated from most parts of the plants especially from pollen, seeds, leaves, shoots and sheaths. The pollen contains remarkably high concentrations, in the case of *Brassica napus* nearly 100 μg brassinolide (**1**)/kg, from *Vicia faba* 190 μg brassinolide (**1**)/kg, and from *Pinus thunbergii* 90 μg 2-deoxycastasterone (**20**)/kg have been found.

Compared to the pollen the content of brassinosteroids in the other parts of the plant normally lies only in the nanogram range. Interestingly, insect galls of *Distylium racemosum* contain 50–300 times more brassinosteroid than the corresponding leaves (2500 ng castasterone (7) and 5000 ng brassinone (**12**) compared to 113 ng and 16 ng, respectively [41]. However, brassinolide (**1**) was found only in the leaves and not in the galls in that case. The reason for such an enrichment has not yet been clarified. However, in the insect galls of *Castanea crenata* the level of brassinosteroids lies in the same order as in the leaves [42].

In the hitherto investigated plant species, brassinolide (**1**) and castasterone (7) have been found to be the most important members, normally accompanied by other brassinosteroids. A broader spectrum of various brassinosteroids was detected in *Dolichos lablab*, *Phaseolus vulgaris*, *Thea sinensis*, *Brassica campestris*, *Distylium racemosum* and *Oryza sativa.*

4 Analysis

The isolation of the quite polar brassinosteroids is normally achieved by extraction the plant material with methanol or isopropanol followed by separation procedures including column chromatography, gel filtration, other partition processes and finally preparative HPLC or TLC. The separation in all cases is monitored by a sensitive bioassay, especially the rice lamina inclination test [61].

For detection and quantification of brassinosteroids GC-MS and GC/CI-MS with computerized selected ion monitoring technique is the method of choice where the corresponding bismethane boronates [17, 62] or methane boronate trimethylsilylethers [62, 63] were found to be the most suitable derivatives [41, 45, 62]. This method has been applied widely for screening and detection of new brassinosteroids up to picogram amounts [15, 19, 22, 24, 36–38, 42, 46, 54, 55, 58, 64]. The derivatives used exhibit moderate volatility, give the corresponding molecular ions and show characteristic fissions between C23–C24, C20–C22, C17–C20 and for the B-ring lactone as shown in Fig. 1. In the CI-MS the molecular ions of the brassinosteroid derivatives usually represent the base peak [41]. In contrast to the EI-MS the negative ion mass spectra show strong peaks in the molecular ion region, too. Interestingly, besides a $[M\text{-}1]^-$ ion the brassinosteroids display a $[M\text{-}4]^-$ ion typical of the vicinal diol function in the side chain [65].

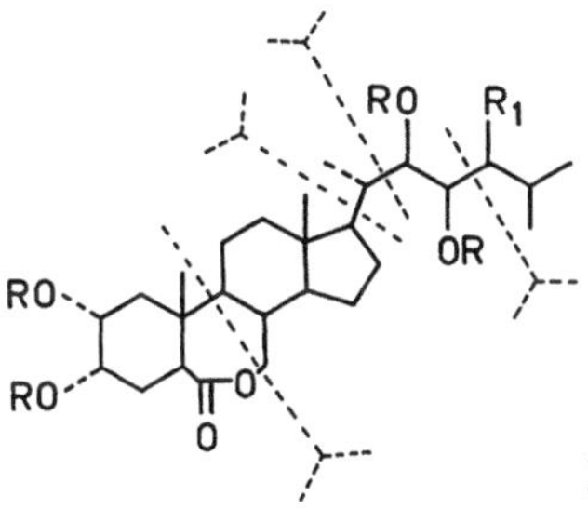

Fig. 1

Table 2. Occurrence of brassinosteroids

Plant	Part of plant	1	2	3	4	5	6	7	11	12	13	14	15	17	20	21	24	25	26	28	References
1. *Alnus glutinosa* L. (European Alder)	pollen	+						+													[43]
2. Beeswax		+																			[44]
3. *Brassica campestris* var. pekinensis Lour.	seed	+	+	+		+		+	+	+											[17, 41, 45]
	sheath	+	+					+													[17, 45]
4. *Brassica napus* L.	pollen	+																			[1]
5. *Castanea crenata Sieb* et Zucc.	galls	+		+		+		+	+	+							+				[28, 41, 42, 46]
6. Castanea spp.	galls	+						+													[45]
7. *Catharanthus roseus* (L.) G. Don	galls	+						+													[47]
8. *Chrysanthemum golden* rod s. altissima								+													[48]
9. *Diospyra kaki* Thunb.								+													[49]
10. *Distylium racemosum* Sieb et Zucc.	galls	+		+		+		+	+	+											[41]
	leaves	+		+		+		+	+	+											[41]
11. *Dolichos lablab* L.	seed	+			+	+		+			+	+					+	+			[21, 22, 24, 50]

12. *Equisetum arvense* L.	strobilus		+			+		+	+									[51]
13. *Fagopyrum esculentum* Moench	pollen	+				+												[52]
14. *Helianthus annuus* L.	pollen	+				+		+										[53]
15. *Hydrodictyon reticulatum*							+			+								[30]
16. *Oryza sativa* L. cv. Arborio J 1	plant	+	+	+		+	+	+										[41]
17. *Pharbitis purpurea* Roth	seed					+		+										[54]
18. *Phaseolus vulgaris* L. cv. Kentucky Wonder	seed	+							+		+			+	+	+	+	[12, 13, 14] [37, 38]
19. *Picea sitchensis* Bong.	shoots					+						+						[55]
20. *Pinus thunbergii* Parl.	pollen											+						[56]
21. *Raphanus sativus* L.	seed	+				+												[57]
22. *Thea sinensis* L.	leaves	+	+	+		+	+	+				+	+					[41, 45, 58]
23. *Typha latifolia* L.	pollen											+						[35]
24. *Vicia faba* L.	seed	+				+												[59, 60]
	pollen	+			+	+		+										[26]
25. *Zea mays* L.	pollen					+						+	+					36

Microanalytical methods using fluorimetric detection of the dansyl aminophenyl boronates [66] or the phenanthrene boronate derivatives [67, 68] have been described, too. In those cases the detection limit was found to be 25 pg and 100 pg, respectively, for brassinolide (**1**).

Monoclonal antibodies directed against 24-*epi*brassinolide (**6**) were produced in CAF_1 mice [69] and an enzyme-linked immunosorbent assay (ELISA) used for the examination of brassinolide distribution in *Brassica napus* tissues was reported [70]. Very recently, a radioimmunoassay (RIA) for brassinosteroids has also been published and has been used successfully for analysing the endogenous content in seeds and stems of *Phaseolus vulgaris*. The detection limits of brassinolide (**1**) and castasterone (**7**) were found to be approximately 0.3 pmol [70a].

Detailed NMR analysis of brassinosteroids have been achieved [70b], and X-ray analysis for brassinolide (**1**) [1], (22*S*,23*S*)-*homo*brassinolide [71], 24-*epi*brassinolide (**6**) and 22,23,24-tri*epi*brassinolide [72] were carried out.

5 Synthesis

5.1 A/B Ring Modification

For the chemical synthesis of brassinosteroids starting from a suitable precursor, the typical A/B ring functionalization and the construction of the dihydroxylated side chain moiety including the asymmetric centres at C-20, C-22, C-23 and C-24 are necessary. Concerning the introduction of the characteristic structural functions in rings A/B 6-ketones of typ **30** were important intermediates, for which the following pathways were developed:

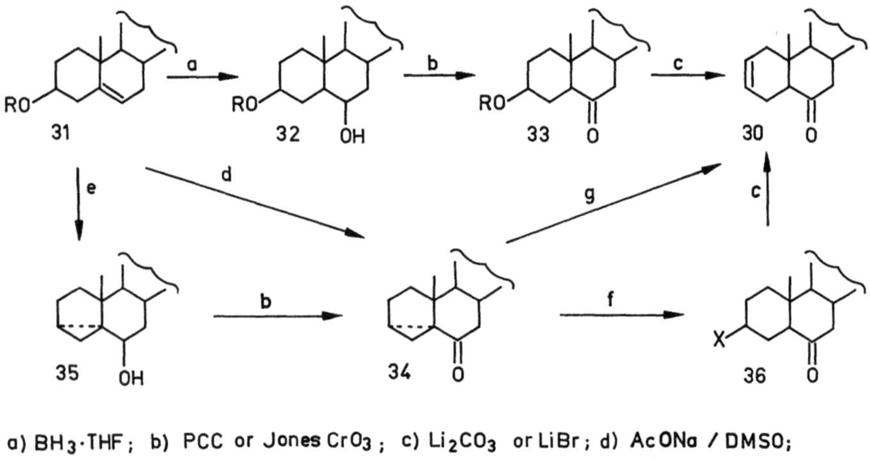

a) $BH_3 \cdot THF$; b) PCC or Jones CrO_3; c) Li_2CO_3 or LiBr; d) AcONa / DMSO; e) $KHCO_3$/acetone; f) H_2SO_4/AcOH or HCl/AcOH or HBr/AcOH; g) p-TsOH/sulfolane

– Starting with a 3β-hydroxy-Δ^5-steroid **31** via protection of the hydroxy function and hydroboration the 6-hydroxy intermediate **32** is available, which can be oxidized to the 6-ketone **33**. Elimination with LiBr or Li_2CO_3 in DMF leads to the Δ^2-6-ketone **30** [20, 73, 74].
– Reaction of the 3-mesylate of a 3β-hydroxy-Δ^5-steroid **31** ($R=CH_3-SO_2$) with NaOAc in DMSO gives directly the 3α,5-*cyclo*-6-ketone **34** [75]. An alternative pathway leads via the 3α,5-*cyclo*-6-hydroxy compound **35** and oxidation with CrO_3/pyridine or Jones CrO_3 to **34** [16, 27–29, 39, 76]. From the 3α,5-*cyclo*-6-ketone **34** the Δ^2-6-ketone **30** is available via the 3-chloro, 3-bromo or 3-tosyloxy compound **36** and subsequent elimination [20, 27, 29, 33, 73, 74, 77, 78]. Upon reaction with *p*-toluene sulfonic acid in sulfolane, the direct rearrangement of **34** to **30** is described [16, 39, 76]. Many efforts were made to find more efficient conditions for preparation of the Δ^2-6-ketone **30**, starting from 3α,5-*cyclo* precursors. Depending on the applied reaction conditions, the main product is **30** or the isomeric Δ^3-6-ketone, and it must be added that the Δ^3-6-ketone act as precursor of **30** [79]. The best results were obtained with *p*-TsOH/pyridine/ LiBr [79, 80], *p*-TsOH/NaBr/DMF [81, 82], H_2SO_4/pyridine [83] or HCl/$MgBr_2$ [81]. A further increase in the yield of the desired Δ^2-compound was obtained using a 6β-benzylether [82, 84].
– Using ergosterol (**37**) as starting material, two different pathways to the intermediate 3α,5-*cyclo*-6-ketone **34** have been published. A first sequence includes the isomerization of ergosteryltosylate to the 3α,5-*cyclo*-6-hydroxy-$\Delta^{7,22}$-steroid **38**, whose oxidation to **39** followed by Birch reduction gave the 3α,5-*cyclo*-6-ketone **34** [27]. A second way involves the preferred reduction of the $\Delta^{5,7}$-diene function to the Δ^5-olefine **31** either directly with Li [85–87] or of the 1,4-*cyclo* adduct prepared from 4-phenyl-1,2,4-triazoline-3,5-dione [88] in different solvents.

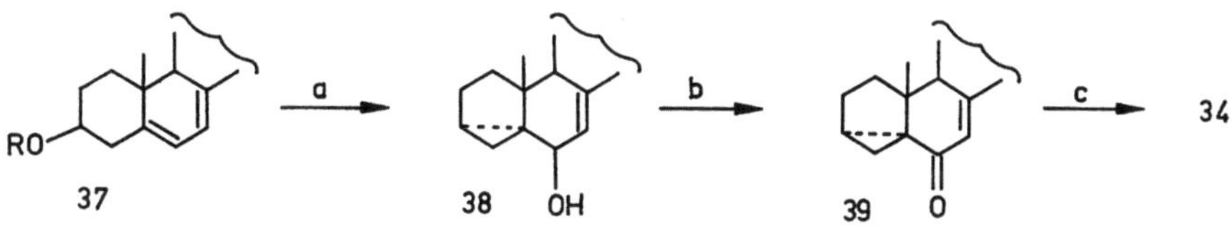

a) K_2CO_3/acetone ; b) CrO_3/pyridine ; c) Li/NH_3

The A/B ring modification is completed by hydroxylation of a Δ^2-6-ketone **30** with OsO_4/NMMNO to **40** and subsequent Baeyer-Villiger oxidation, leading to the 7-oxa-6-oxo-lactone **41** and about 10% of the undesired 6-oxa-7-oxo-lactone **42** [23, 39, 73, 74, 89].

The formation of the 6-oxa-7-oxo-lactone **42** was excluded using the following procedures [23, 90, 91]. The 3α,5-*cyclo*-6-ketone **34** was silylated with trimethylsilyltriflate to give the trimethyl-silyl enolether **43**, which was oxidized with MCPBA to the 6-keto-7-hydroxy compound **44** (as 7-epimers).

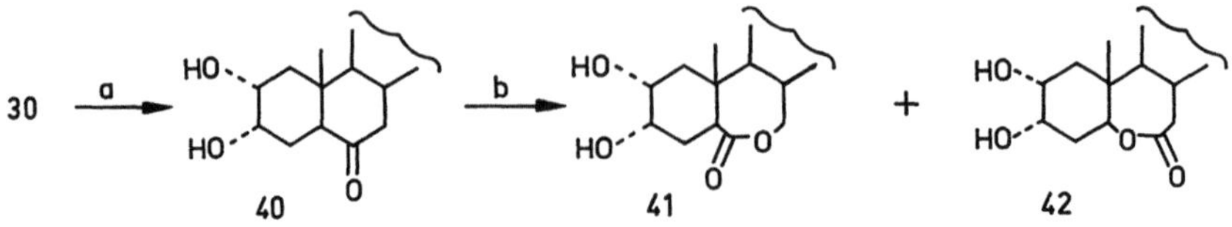

a) OsO_4 / NMMNO ; b) CF_3CO_3H or MCPBA

Rearrangement of **44** with LiBr/ *p*-TsOH afforded the Δ^2-hydroxy-ketone **45**. Ring opening with HIO_4, $NaBH_4$ reduction of the obtained aldehyde and acid catalyzed ring closure leads exclusively to the 7-oxa-6-oxo-lactone **46**.

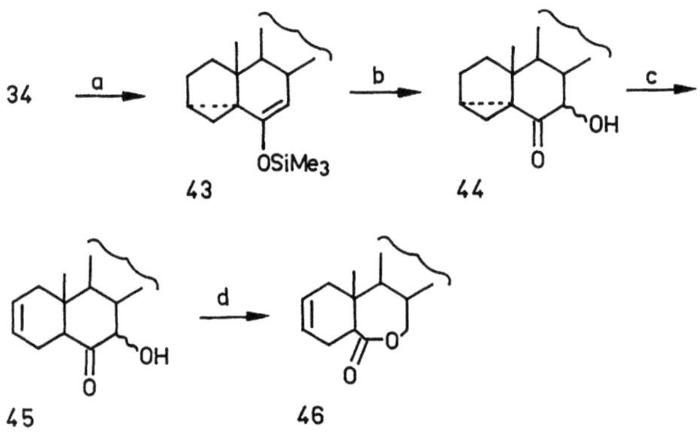

a) $CF_3SO_2SiMe_3$; b) MCPBA ;

c) LiBr ; d) 1. HIO_4, 2. $NaBH_4$, 3. HCl

Another pathway to this intermediate starting from 3β-tosyloxy-6-ketone **33** involved bromination, followed by reaction of the non-isolated 7-bromo compound with aqueous pyridine to give the hydroxy ketone **47**. Oxidative cleavage with HIO_4 and followed by treatment with CrO_3 leads to the 6,7-*seco*-dioic acid **48**. Formation of the intramolecular anhydride **49** and selective reduction with $LiAlH_4$ gave the B-ring lactone **50** [92].

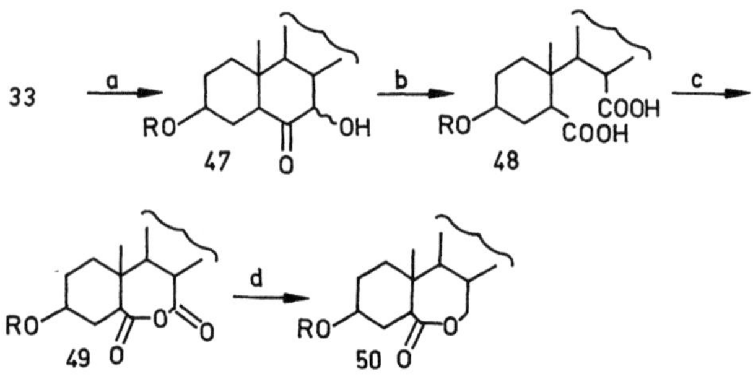

a) 1. Br_2, 2. H_2O/pyridine ; b) 1. HIO_4, 2. CrO_3 ;

c) $(AcO)_2O$; d) $LiAlH_4$

In a more recent reaction sequence to the desired lactone **41** the deoxycholanic acid **51** was oxidized with PDC to the A/B-*cis* fused 3,6-diketone, which via acid catalyzed isomerization yielded the A/B-*trans* fused compound **52**. Reductive elimination of the 3-keto group with Zn/Hg leads to the key intermediate **30**. This method represents one of the most effective ways for constructing the A/B-ring functionality of brassinosteroids [93].

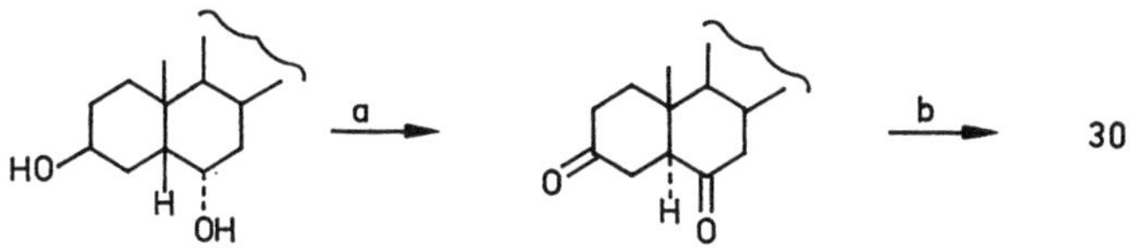

51 52

a) 1. PDC, 2. H^+; b) Zn/Hg

Whereas osmylation of the $\Delta^{2,4}$-diene-6-one **53** gave exclusively the corresponding 2*β*,3*β*-diol **54**, reaction of **53** under Woodward conditions afforded the 2*α*,3*α*-diacetoxy enone **55**, which can be reacted with $LiAlH_4$ to give the 2*α*,3*α*-dihydroxy-6-ketone **40** [94].

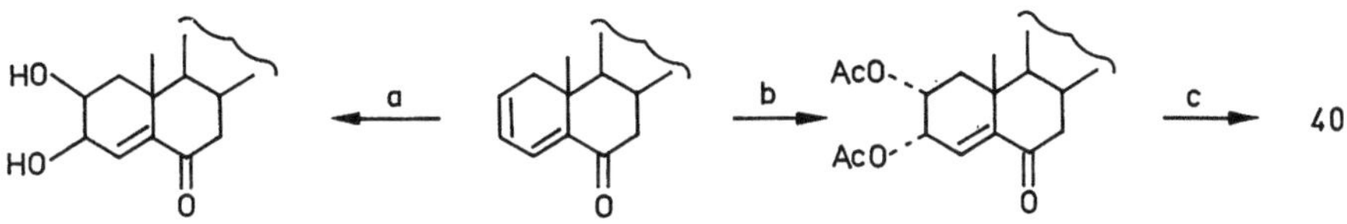

54 53 55

a) OsO_4; b) $Ag(OAc)_2/I_2$; c) $LiAlH_4$

2-Deoxybrassinosteroids like typhasterol (**20**) were synthesized from 3*β*-hydroxy-6-ketone **33** by refluxing its mesylate with Li_2CO_3/DMF. Via inversion, it yields the 3*α*-formiate which can be hydrolyzed to the 3*α*-hydroxy-6-ketone. Another pathway involves the reaction of a 3*β*-halogeno-6-ketone (**36**) with AgOAc followed by hydrolysis of the formed 3*α*-acetoxy intermediate [95–97].

In a synthesis of 6-deoxohomodolichosterone (**25**) from stigmasterol [40] as key step the Birch reduction of the Δ^4-3-ketone **56** gave directly the 3*β*-hydroxy compound **57**. Treatment of the corresponding mesylate with LiBr/DMF and

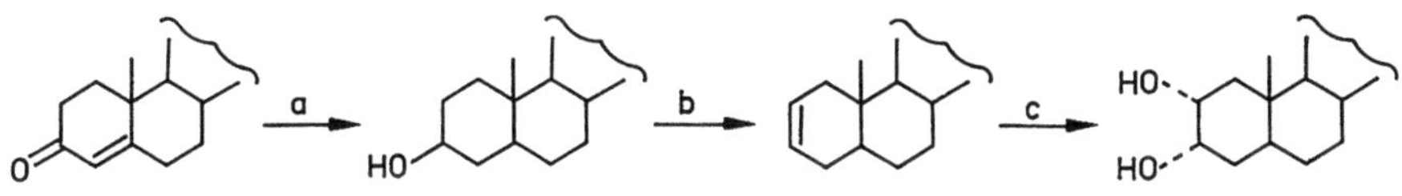

56 57 58 59

a) Li/liq. NH_3; b) 1. MsCl, 2. LiBr; c) OsO_4/NMMNO

hydroxylation of the resulting Δ^2-olefine **58** afforded the desired 6-deoxo diol **59**.

An alternative route to the 2α,3α-diol **40** from Δ^2-unsaturated 5α-sterols **30** avoiding the use of OsO_4 has been described also [98]. It involves the preparation of a 2α,3α-epoxide **60** and cleavage of the oxirane ring with HI to give the iodoalcohol **61**. Acetylation, substitution of the iodine by hydroxyl and hydrolysis provides the desired 2α,3α-diol **40**.

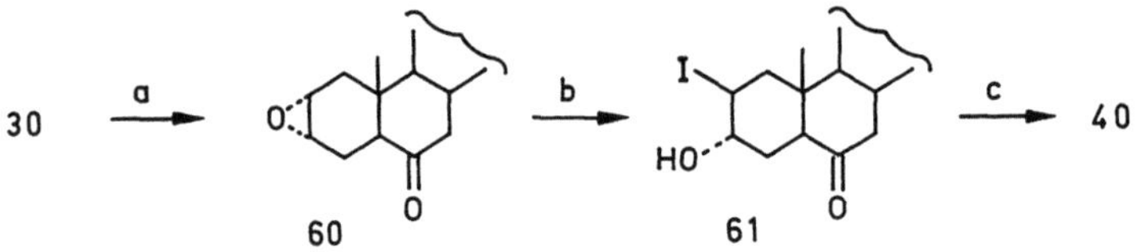

a) RCO_3H; b) HI; c) 1.$(AcO)_2O$, 2.RCO_3H, 3.OH^-

5.2 Side Chain Construction

As important starting materials with a preformed side chain, Δ^{22}-sterols **62** such as 22,23-didehydrocampesterol (R = (24*S*)-Me), brassicasterol (R = (24*R*)-Me), stigmasterol (R = (24*S*)-Et), ergosterol (R = (24*R*)-Me), 22-dehydrocholesterol (R = H) and poriferasterol (R = (24R)-Et) have been used for brassinosteroid synthesis. The introduction of the vicinal 22,23-diol function was effected either upon direct hydroxylation or via epoxidation followed by epoxide fission. However, in all cases a mixture of the stereoisomeric 22,23-diols **63** and **64** or epoxides **65** and **66** were obtained, respectively. The ratio of the epimers is strongly influenced by the stereochemistry and the type of substituents at C-24. These earlier synthetic pathways are discussed in detail in Ref. [3].

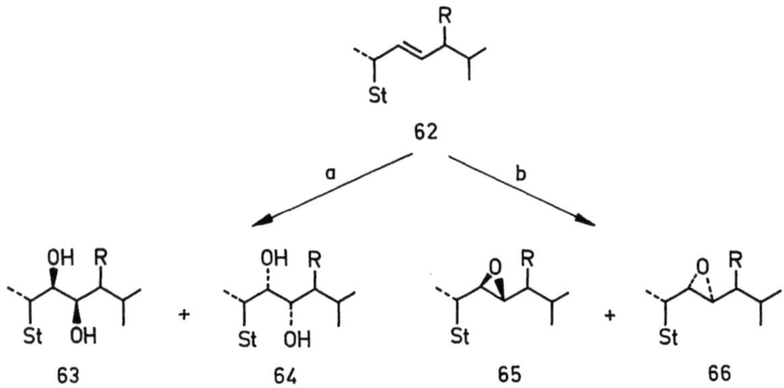

a) OsO_4 /NMMNO; b) MCPBA

Especially in the earlier syntheses of natural brassinosteroids, as preferred starting material 22-aldehydes of type **67** were used for side chain constructing, readily available from dinorcholanic acid [99] or stigmasterol [100–102]. The different sequences are summarized in the following scheme.

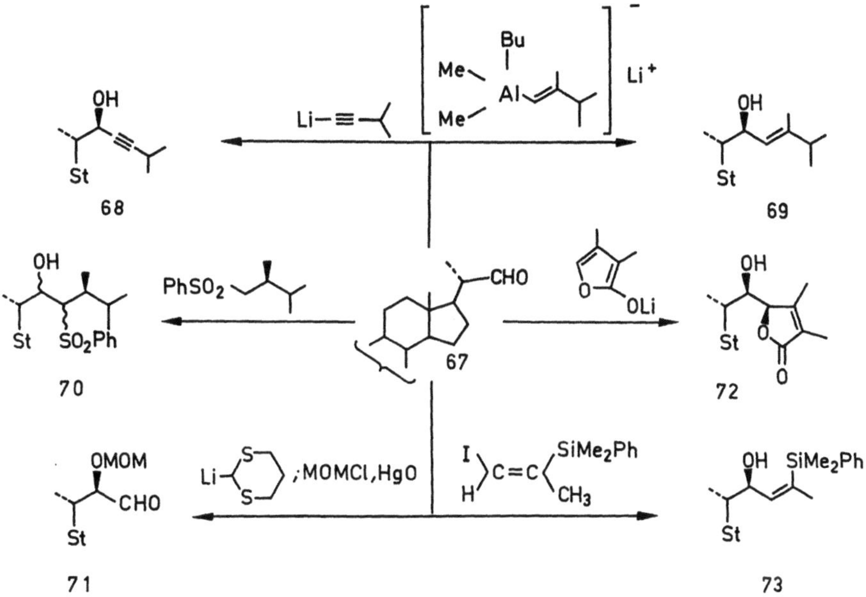

Thus, starting from **67**, Ishiguro et al. [73, 89] used 3-methylbut-1-inyl-lithium to synthesize the key inylalcohol **68**. Similarly, Fung and Siddall [74] applied $[Me_2BuAlCH=CMeCHMe_2]Li$ as organometallic reagent to get the allylalcohol **69**. Later similar alkylating reagents were used by other authors [23, 29, 32, 89, 102]. In another synthesis published by Mori et al. [16, 39], the (2*R*)-(+)-(2,3-dimethyl)butanyl phenyl sulfone as a chiral building block was reacted with **67** to give phenyl sulfone **70**.

A quite variable method for the construction of the brassinosteroid side chain was developed by Takatsuto and Ikekawa [63, 96, 103, 104] using the chelation-controlled Grignard reaction for a stereoselective introduction of the desired (22*R*,23*R*)-diol function. Thus, reaction of the 22-aldehyde **67** with 1,3-dithiane lithium, protection of the obtained 22-hydroxy function as MOM-ether followed by desulfuration with HgO yielded the 23-aldehyde **71**, which was transformed with a variety of Grignard reagents to yield the corresponding (22*R*,23*R*)-diols.

The synthesis published by Donaubauer et al. [105] involves the aldol reaction of the 22-aldehyde **67** with the lithium salt of 2,3-dimethyl butenolide to give the preferred (22*R*)-hydroxy compound **72**. In a modified procedure the lithium salt of 3-isopropyl butenolide was used.

A further synthetic pathway was developed by Hayami et al. [106]. The key step is the reaction of **67** with the lithium salt of 2-dimethylphenylsilyl-1-iodopropene to give **73**.

More recently, Kametani et al. [107] published a stereocontrolled construction of the brassinosteroid side chain, using 3-isopropyl-2-lithiofuran as building block. Starting from **67**, the yield of the desired (22*R*)-alcohol **74** was 81% besides 18% of the (22*S*)-epimer. Oxidative ring expansion with MCPBA to **75** followed by oxidation with PCC led to the diketone **76**. Hydrogenation over PtO_2, reduction with $NaBH_4$ and alkaline treatment gave predominantly the (22*R*)-hydroxylactone **77** under ring contraction. Reductive ring opening with $LiAlH_4$ and reductive elimination of the 26-hydroxy function via the 26-mesylate completed the brassinolide side chain construction [108]. In a modified procedure the endione **76** was reduced with $NaBH_4$ to the allylic alcohol **78** and further hydrogenated over PtO_2 to give **77** in a higher overall yield.

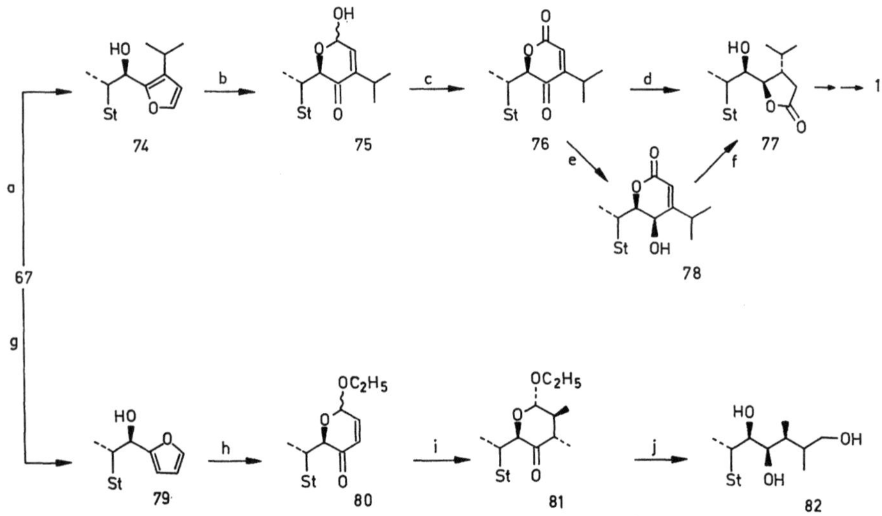

a) 3-iProp-2-Li-furan; b) MCPBA; c) PCC; d) 1. H_2/PtO_2, 2. $NaBH_4$, 3. OH^-
e) $NaBH_4$; f) H_2/PtO_2 g) 2-Li-furan; h) NBS; i) 1. $LiMe_2Cu$ 2. MeI/LDA;
j) $LiAlH_4$

In a similar reaction sequence, aldehyde **67** was transformed with 2-lithiofuran to give predominantly the (22*R*)-furylcarbinol **79** [109]. The conversion of the minor (22*S*)-alcohol to the (22*R*)-epimer was achieved via an oxidation/reduction sequence. Reaction of the furan with NBS gave a mixture of epimeric alcohols, which were converted to the ethoxy ethyl ethers **80**. Reaction of the separated main product with lithium dimethylcuprate and alkylation with MeI/LDA afforded **81**, whose reduction with $NaBH_4$ followed by lactol formation and reaction with $LiAlH_4$ yielded the triol **82**.

In a modification, the ring expansion starting from **79** was realized with MCPBA to give **83** [110]. Silylation and reduction with Dibal-H yielded **84, 85** and **86** in a 1:1,6:1 ratio. Benzylation, desilylation and oxidation of **84** and **85** led to the lactone **87**, which was further transformed into the 28-*nor*brassinolide (**3**) as described earlier [109].

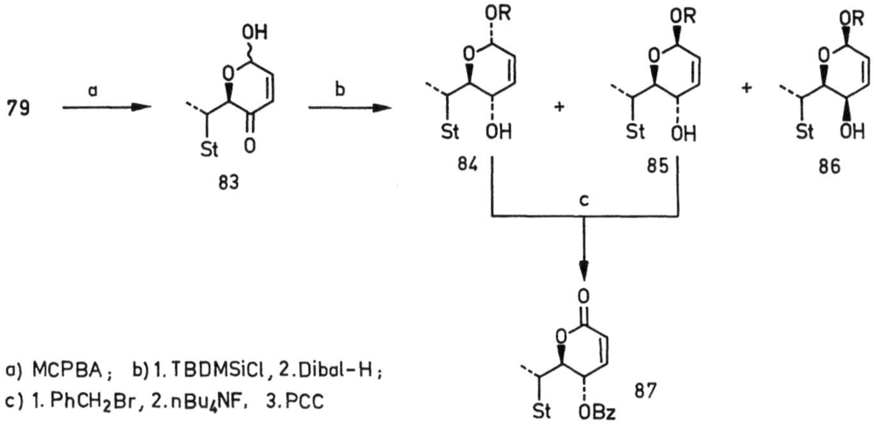

In a similar synthesis [111] the aldehyde **67**, via reaction with 3-methyl butenolide and LDA, afforded the Cram product **88**. Hydrogenation of **88** over PtO_2 and methylation with MeI/LDA gave **89**, whose reduction with $LiAlH_4$ followed by elimination of the 26-hydroxy function yielded the brassinolide side chain. An increase in the yield could be obtained by hydrogenation of the acetate **90** and

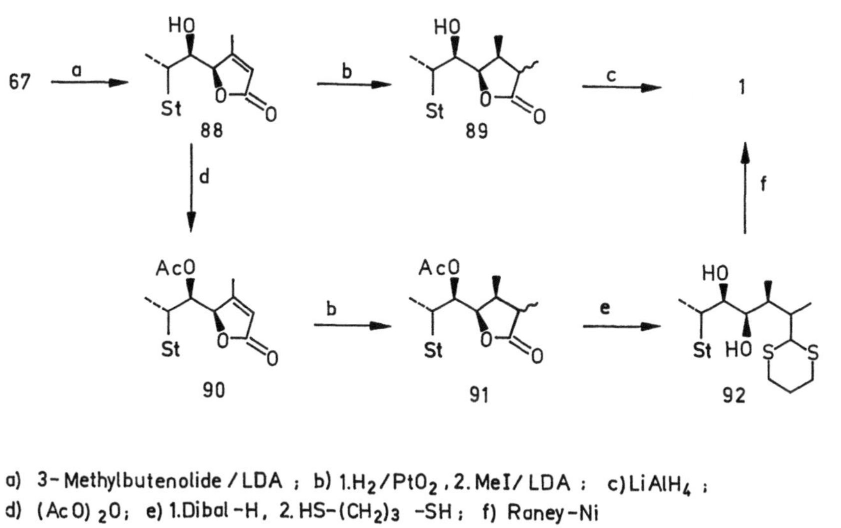

alkylation to **91**. Reduction with Dibal-H, conversion with the 1,3-dithiol to give **92** and reduction with Raney nickel completed the synthesis.

A new principle for side chain construction was published by Zhou et al. [93, 112, 112a] with the reaction of the aldehyde **67** with the arsonium salt, prepared from Ph_3As and 1-bromo-3-methyl-2-butanone. Epoxidation of the resulting Δ^{22}-24-ketone **93** to **94** and Wittig-Horner reaction with ethoxycarbonylmethyl phosphonic acid dimethylester led to the (*E*)- and (*Z*)-epoxyolefins **95** in a 10:1 ratio. Acid catalyzed lactonization and inversion of the simultaneously formed hydroxy function at C-23 via an oxidation/reduction sequence completed the preparation of the hydroxylactone **78**, which was transformed into brassinolide in the known manner.

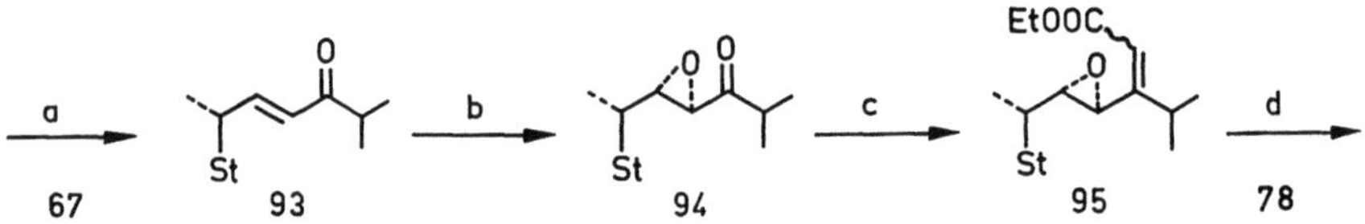

a) $Me_2CH_2COCH{=}AsPh_3$, b) H_2O_2, c) $(MeO)_2P(O)CH_2CO_2Et$,
d) 1. $HClO_4$, 2. PDC, 3. H_2/PtO_2

The construction of the side chain portion of brassinosteroids from the 22-aldehyde **67** using metalacetylides normally leads to mixtures of the Cram and the anti-Cram stereoisomers **96** and **97**, respectively, in a 3:2 ratio [29]. However, Yamamoto et al. [113] showed, that the reaction of **67** with stannylacetylenes, allylstannanes and allylsilanes in the presence of $TiCl_4$ gave predominantly the desired Cram isomer **96**.

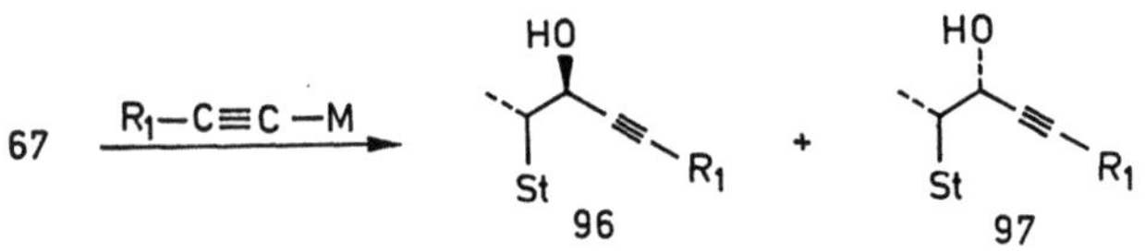

Back et al. [114] developed a selenosulfonation-based approach to the 23-en-22-ol **69** as key intermediate for brassinolide synthesis. Thus, aldehyde **67** was treated with 3-lithio-1-trimethyl-silylpropyne to afford after desilylation a 1,3:1 mixture of epimeric C-22 alcohols **98**. The mixture was subjected to free radical selenosulfonation with Se-phenyl p-toluene selenosulphonate to afford the product of *anti* addition **99**, which was separated chromatographically. Base-catalyzed isomerization produced the allylic sulfone **100**. Oxidation and selenoxide *syn* elimination, induced with t-butylhydroperoxide, gave the allenic sulfone **101**, which was treated with lithio-diisopropyl cuprate · SMe_2 to afford the allylic sulfone **102**. The best results in the final reductive desulfonylation to the allylic alcohol **69** were obtained with magnesium in methanol.

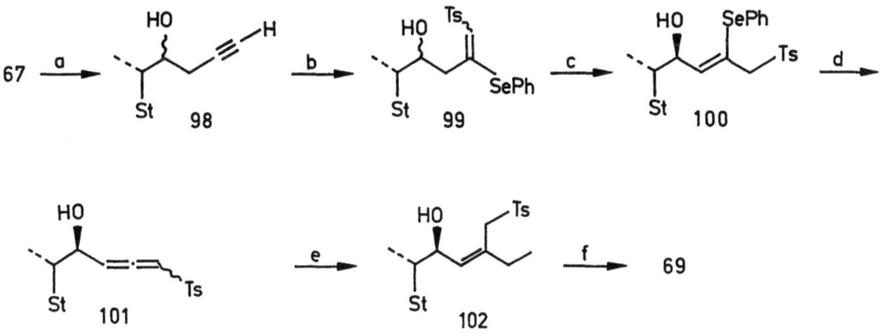

a) 1. $LiCH_2C{\equiv}CSiMe_3$, 2. nBu_4NF; b) TsSePh; c) NEt_3;
d) tBuOOH; e) $LiCu(iProp)_2 \cdot SMe_2$; f) Mg/MeOH

Soon after the isolation of the highly active 25-methyldolichosterone (**17**) [13] the following synthesis for 25-methylbrassinosteroids has been published [34]. Thus, reaction of the 22-aldehyde **67** with the lithium salt prepared from t-butyl acetylene/$Me_3Al/ZrOCl_2$ and butyllithium led to the allylalcohol **103**. Epoxidation to **104** and rearrangement with $(iso\text{-}PropO)_3Al$ gave 25-methyldolichosterone (**17**). Catalytic hydrogenation and Baeyer Villiger oxidation yielded 62% 25-methylbrassinolide (**105**) besides 10% 24-*epi*-25-methylbrassinolide (**106**).

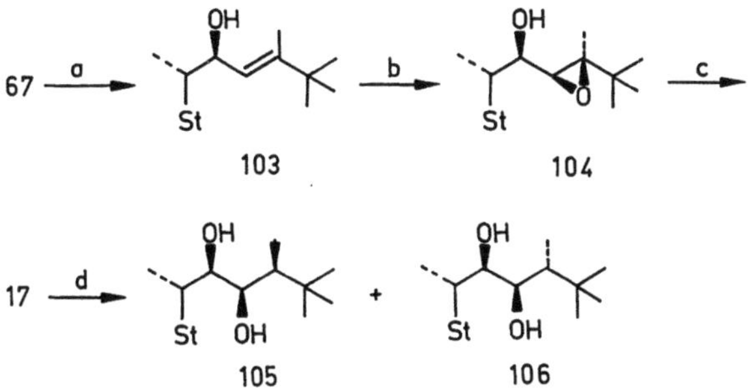

a) $tBuC{\equiv}CH/Me_3Al/ZrOCl_2/LiBu$; b) MCPBA
c) $(iPropO)_3Al$; d) H_2/Pd-C

A pathway to brassinosteroid key intermediate **69**, starting from the tosylate **107**, readily available from 3β-hydroxypregn-5-en-20-one, was described by Tsuji et al. [115]. Thus, alkylation with a protected C_8-cyanhydrin building block **108** followed by acid and alkaline treatment gave the enone intermediate **109**, which was reduced with Dibal-H to the (22*R*)-allylic alcohol **69**, a known intermediate in the brassinolide synthesis of Mori et al. [23].

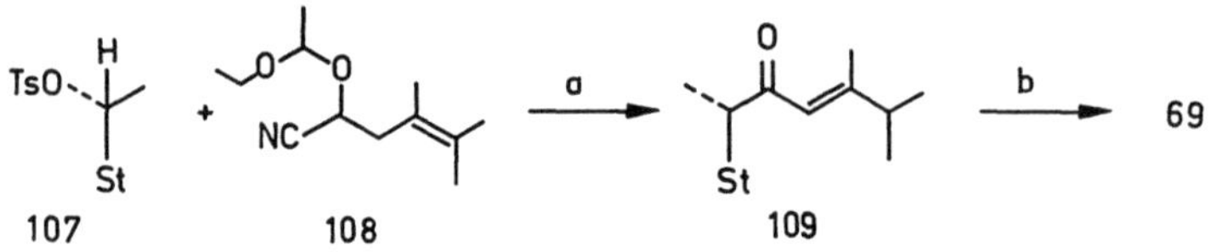

a) 1 H^+, 2 OH^-; b) Dibal-H

An improved synthesis of brassinolide and its congeners starting with 20-ketopregnane **110** was also published [108, 116]. Reaction of **110** with 3-isopropyl tetronic acid (R_1 = iso-Prop) in the presence of LDA gave the 20-hydroxy lactone **111** followed by dehydration to the olefin **112** and hydrogenation over 5%

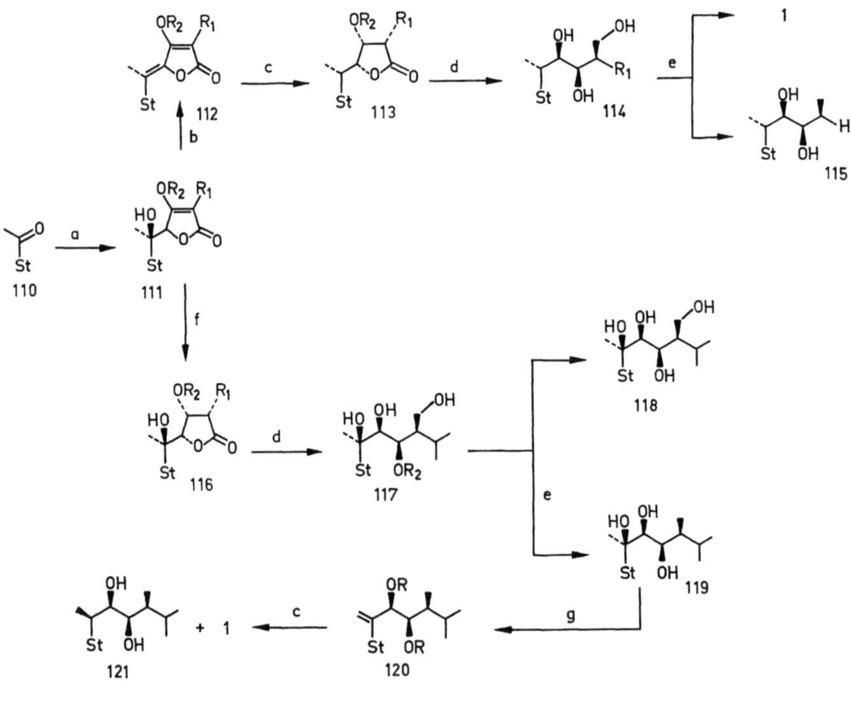

a) 3-iProp tetronic acid/LDA; b) $(CF_3CO)_2O$; c) $H_2/Rh/Al_2O_3$; d) $LiAlH_4$
e) 1. MsCl, 2. $LiAlH_4$; f) $H_2/Rh/Al_2O_3$/ethyl acetate; g) 1. $(AcO)_2O$, 2. $(CF_3CO)_2O$

Rh/Al_2O_3 to yield the saturated lactone **113**. The side chain synthesis was completed by $LiAlH_4$ reduction to the triol **114** and by reductive elimination of the primary alcohol via its mesylate to give **1**. In the case of tetronic acid ($R_1 = H$) as starting reagent the 26,27-di*nor*brassinolide side chain **115** was obtained. Hydrogenation of the unsaturated lactone **111** over Rh/Al_2O_3 yielded the saturated lactone **116**, which was transformed via **117** into 20,28-dihydroxybrassinolide (**118**) and 20-hydroxybrassinolide (**119**), respectively. After acetylation of the secondary hydroxy functions of **119** and dehydration the resulting $\Delta^{20(21)}$-olefine **120** was hydrogenated and hydrolyzed to give brassinolide (**1**) and 20-*epi*brassinolide (**121**) in equal amounts [117].

A stereocontrolled pathway either to (22*S*)- or to (22*R*)-hydroxy steroid intermediates starting from $\Delta^{17(20)}$-olefins was published by Mikami et al. [118]. Thus, reaction with an inylaldehyde in the presence of Me_2AlCl afforded predominantly the (22*R*)-hydroxy steroid **96** besides the epimeric **97**. Further ways to other side chain intermediates starting from $\Delta^{17(20)}$-pregnenes have been published recently [51, 119, 120].

5.3 Labelled Compounds

Labelled brassinosteroids are important for investigations of biosynthesis, metabolism, transport, distribution and mode of action of these phytohormones, where only few studies have been done in this direction till now. As a suitable precursor for first deuterated brassinosteroids [26,28-$^2H_6$]crinosterol (**124**) was synthesized from the aldehyde **67** [121]. Thus, reaction with lithium acetylide gave predominantly the (22*R*)-alcohol **96**. Alkylation with [$^2H_3$]iodomethane and partial hydrogenation yielded the allyl alcohol **122**, which was submitted to an orthoester Claisen rearrangement reaction to give the ester **123**. The ethoxycarbonyl group was transformed into the [$^2H_3$]methyl group via $LiAl^2H_4$ reduction, mesylation and again $LiAl^2H_4$ reduction. The resulting [26,28-$^2H_6$]crinosterol (**124**) was converted in the known manner into [26,28-$^2H_6$]brassinolide, [26,28-$^2H_6$]typhasterol and [26,28-$^2H_6$]teasterone [122, 123].

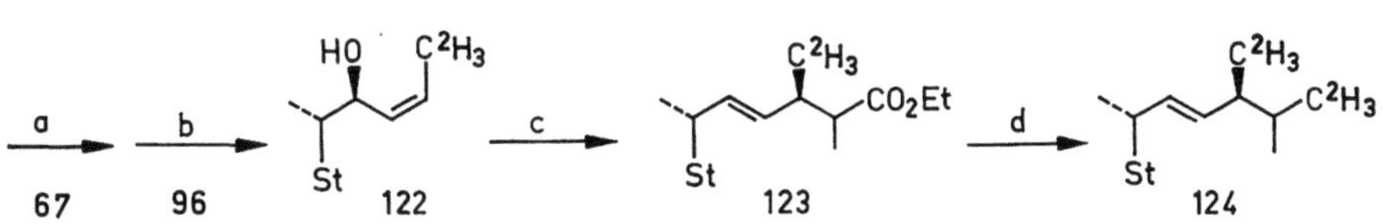

a) HC≡CLi; b) 1.^2H_3CI, 2.H_2/PtO_2; c) Claisen rearrangement;
d) 1.$LiAl^3H_4$ 2.MsCl 3.$LiAl^3H_4$

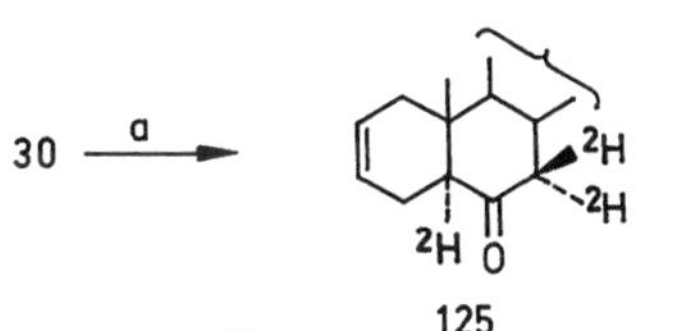

a) Na/MeO^2H

Deutero-labelled 24-*epi*brassinolide (**6**) was prepared starting from ergosterol based on the synthesis of the corresponding unlabelled compound [240, 243]. Thus, intermediate Δ^2-6-ketone **30** was refluxed with Na/MeO^{2}H and hydrolyzed with ^2H_2O. The resulting [5,7,7-$^2H_3$]-Δ^2-6-ketone **125** was transformed into [5,7,7-$^2H_3$]-24-*epi*brassinolide, used as standard for studies of its uptake in growing maize root segments [124].

A pathway to first ^{14}C-labelled brassinosteroids was reported recently [125]. Thus, the enelactone **126**, available from brassicasterol in 5 steps via the Grignard reaction with [^{14}C]methyliodid gave the bridged ketone **127**, which upon alkaline treatment furnished [4-^{14}C]brassicasterone (**128**). Enolacetylation of **128** with isopropenyl acetate and NaBH$_4$-reduction afforded [4-^{14}C]brassicasterol (**129**), which was converted to [4-^{14}C]-24-*epi*brassinolide in the convential manner.

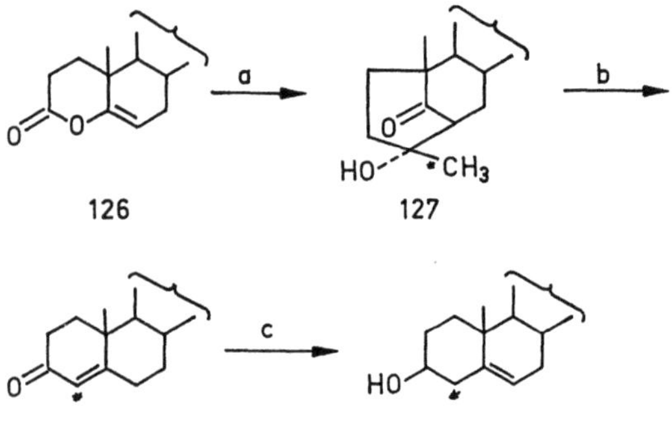

126 127

128 129

a) $^{14}CH_3MgI$; b) pTsOH; c) 1. NaBH$_4$ 2. AcOH

Very recently, [24,28-$^3H_2$]castasterone and its 24-epimer were prepared from dolichosterone (**13**) by catalytically tritiation of the $\Delta^{24(28)}$-double bond over platinum. Likewise, dolicholide (**4**) was converted into [24,28-$^3H_2$]-brassinolide (**130**) and [24,28-$^3H_2$]-24-*epi*brassinolide (**131**). Both obtained radio-labelled compounds have been used for a radioimmunoassay of brassinosteroids [70a].

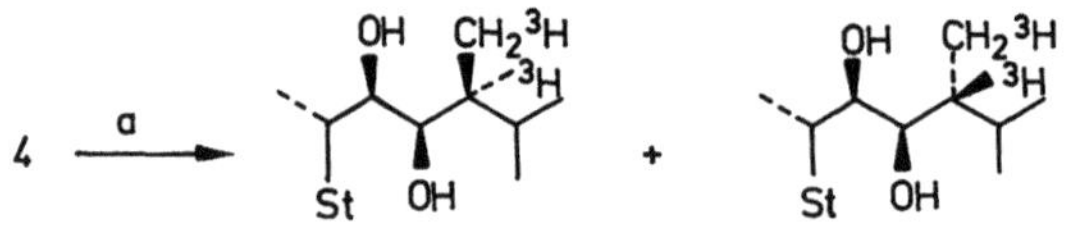

130 131

a) $^3H_2$ / PtO$_2$

Feeding experiments with [24,28-$^3H_2$]castasterone on crown gall cells of *Catharanthus roseus* brought the evidence for the first time that **7** is a biosynthetic precursor of brassinolide (**1**) [126].

5.4 Structural Analogs

For structure-activity relationships a large number of structural analogs of brassinosteroids have been synthesized [3]. Among the side chain modified compounds are epimers with (22*S*,23*S*)-, (22*R*,23*S*)- or (22*S*,23*R*)-dihydroxy function, members with a different number (up to 3) or position of hydroxy functions, side-chain-shortened compounds as well as aza and thio analogs, which are summarized in Ref. [3].

Recently, prepared side chain analogs have been synthesized starting with androstane or pregnane compounds [127–129] as well as diosgenin and other spirostans [130, 131].

The preparation of A/B ring analogs, including alkyl-substituted members [132], compounds with hydroxy functions in different positions [132, 133], unsaturated B-ring [94, 133, 134], halogeno-substituted members [135] as well as 2,3-*seco*- [135a] and 5,6-*seco*-brassinosteroids [130] has also been published.

Dealing with the microbial 12β-hydroxylation of (22*S*,23*S*)-*homo*brassinolide and (22*S*,23*S*)-*homo*castasterone using *Cunninghamella echinolata*, the first microbial transformations of brassinosteroids have been reported [136].

6 Biological Effects

6.1 Bioassay

Brassinosteroids were tested for their plant growth promoting activity in more than 20 bioassays typical for auxin, gibberellin or cytokinin activity. In several systems brassinosteroids interact strongly with auxins in a synergistic manner. On the other hand, the response of brassinosteroids and gibberellins appears to be both independent and additive. In systems designed as characteristic for cytokinins, brassinosteroids work in various ways. Thus, brassinosteroids can act auxin-like at one time and gibberellin- or cytokinin-like at an other time. Sometimes, brassinosteroids in the corresponding test systems show a higher activity at lower concentrations than auxins, gibberellins and cytokinins. Most investigations in this field were done by 1986 and have been summarized previously [7–9, 137, 138]. Table 3 shows the most important hitherto used bioassays [6].

From the different assays desired, the rice lamina inclination test, modified by Arima [42], and the bean second internode test [140] represent the most specific bioassays for brassinosteroids and are widely used for detection of their activity. Also the bean first internode curvature bioassay, modified by Strnad [145], as well as the whole plant tests with tomato or *Raphanus sativus* are useful. The wheat leaf unrolling test can also be applied for detection of anti-brassinosteroid activity [143].

Table 3. Bioassays for brassinosteroids

Test	Sensitivity	Ref.
Rice lamina inclination test	1 pmol · l^{-1}	[61]
Bean first internode curvature bioassay	0.1 pmol · l^{-1}	[139]
Bean second internode bioassay	0.1 pmol · l^{-1}	[140]
Maize mesocotyl bioassay	1 μmol · l^{-1}	[141]
Azuki bean epicotyl bioassay	1 μmol · l^{-1}	[141]
Increase in fresh weight of Jerusalem artichoke tissue	10 μmol · l^{-1}	[141]
Multiple hanging drop assay with pollen tubes of *Pinus avium*	1 pmol · l^{-1}	[142]
Wheat leaf unrolling test	1 pmol · l^{-1}	[143]
Dwarf pea epicotyl bioassay	0.1 μmol · l^{-1}	[137]
Inhibition of betacyanin accumulation in *Amaranthus retroflexus* L.	10 μmol · l^{-1}	[137]
Raphanus sativus test	0.001 ppm	[144]
Tomato test	0.01 ppm	[144]

6.2 Structure-Activity Relationships

Based on the results of the rice lamina, bean first and bean second internode bioassays, brassinolide (**1**) shows the highest activity. Comparable high activity in selected bioassays was obtained by some other natural brassinosteroids such as 28-*homo*brassinolide (**2**), 28-*nor*brassinolide (**3**) and 25-methyldolichosterone (**17**), and by synthetic analogs like 26,27-di*nor*brassinolide. All other natural brassinosteroids and synthesized analogs show reduced biological activity, e.g., dolicholide (**4**), 24-*epi*brassinolide (**6**) or castasterone (**7**) approximately 10% of **1**. The loss of more than one of the typical brassinosteroid functionalities leads to a further reduction of the biological activity.

As published by Ikekawa [146, 146a], structural requirements for a high brassinosteroid activity are assumed to be the following:

- (22*R*,23*R*)-vicinal diol moiety
- (24*S*)-methyl or -ethyl group
- 7-oxalactone or 6-oxo functionality in the B-ring
- 3-hydroxy group, 2α,3α-vicinal diol or 3α,4α vicinal diol and
- A/B-*trans*-fused ring function.

However, also 25-methylbrassinolide, 24-*epi*brassinolide (**6**), 22,23,24-tri*epi*-brassinolide and (22*S*,23*S*)-*homo*brassinolide still possess remarkable activity. The last three compounds are used preferably for biological experiments. Reports of other bioactive analogs with modified side chain functions – e.g., carboxylic acids [147] or with spirostan side chain moiety have been published, however, further investigation is necessary. A 12β-hydroxy function lowers the brassinosteroid activity [130, 131].

Surprisingly, the synthesized (22*R*, 23*R*)-epoxy intermediate **132** and the phenyl derivatives of type **133** show a low activity in the bioassay. However, they exhibit a comparable or even higher activity than brassinolide (**1**) in field studies [148–150]. For an explanation of this effect a delayed in-vivo transformation of **132** into an active compound, or in the case of **133** a delayed side chain degradation, can be assumed, suggesting a further search for specific analogs with favourable in-vivo properties.

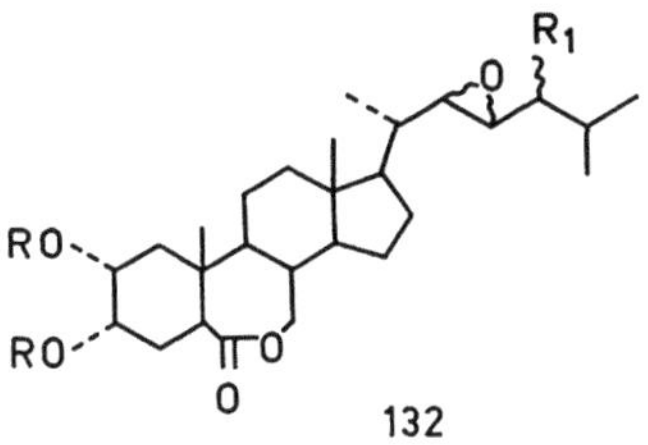

132

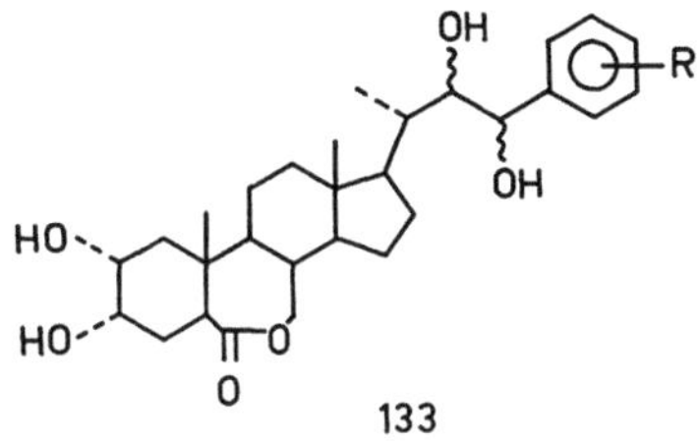

133

6.3 Physiology

The bioactivity of brassinosteroids is characterized by strong growth stimulation including cell elongation and cell division especially in young growing tissues [8, 9, 10]. Similar to the classical phytohormones they show multiple effects on growth and development. Efforts made by several laboratories involved further studies to get deeper insight into the physiology and mode of action.

Thus, brassinosteroids increase the ethylene production by activation of ACC synthesis [151–154]. The synthesis of ACC and the capability of tissue to conjugate ACC to 1-(malonylamino)-*cyclo*propane-1-carboxylic acid (MACC) were stimulated without impact on the ethylene forming enzyme [155]. When brassinosteroids and IAA were applied in combination, a synergistic increase of ethylene production was found. Aminooxyacetic acid, cycloheximide and $CoCl_2$ inhibited the induced ethylene and ACC production [156]. Also 1 µmol fusicoccin inhibited brassinosteroid induced ethylene, ACC and ACC-synthase production [157]. The ethylene formation in dark grown mung bean seedlings after exposition to fluorescent light decreased under the influence of brassinosteroids. Pretreatment with irradiation exhibited a maximum inhibition [158]. Otherwise, there was no influence on ethylene production after brassinosteroid treatment of hypocotyls of etiolated squash (*Cucurbita maxima*) [159]. A side effect caused by stimulated ACC and ethylene production is epinasty, observed after application on hydroponically grown tomato plants [160, 161].

Further investigations were done to study the observed hyperpolarization of the transmembrane electrical potential combined with H^+-extrusion in more detail. The effect is caused by brassinosteroid application, although a correlation between these physiological effects and structure is not yet clear [162–165]. Brassinolide (**1**) stimulates growth and H^+-extrusion in Azuki bean epicotyls and hyperpolarizes the transmembrane electrical potential similarly to IAA. These effects are suppressed by inhibitors of RNA and protein synthesis, but appear roughly additive to IAA [166]. Using maize root segments, the application of brassinolide (**1**) and also of some modified brassinosteroids yields in a growth response and stimulation of H^+-secretion. Characteristically, a lack of the hydroxy groups in the side chain or a change of the 7-oxalactone to the 6-oxalactone type suppress the activity [167]. Stigmasterol, ergosterol and cholesterol, only, stimulate the H^+-extrusion, but are ineffective on growth [168]. The growth of isolated radish cotyledons is much more stimulated by benzyladenine than by 24-*epi*brassinolide (**6**). Both effects were additive at optimal concentrations suggesting different modes of action. Contrary to the brassinosteroid, benzyladenine did not activate H^+-secretion [169]. Applied to *Egeria densa*, (22*S*,23*S*)-*homo*brassinolide was similar to fusicoccin in causing hyperpolarization and H^+-extrusion. In darkness, ferricyanide reduction was stimulated and a promotion of the uptake of sucrose and α-amino*iso*butyric acid was observed [170]. In *Commelina communis*, (22*S*,23*S*)-*homo*brassinolide as well as 24-*epi*brassinolide (**6**) caused inhibition of stomatal opening [171].

Seeking possible relationships between brassinosteroid effects on plant growth and enzyme activity, brassinosteroid induced enhancement of RNA and DNA polymerase activity in isolated nuclei of bean tissue sections has been observed. Also levels of RNA, DNA and protein were increased [172, 173]. On mung bean epicotyls, brassinosteroids interact strongly with putative inhibitors of ribonucleic acid and protein synthesis and can overcome their effects on growth [174]. The influence of 24-*epi*brassinolide (**6**) on the activities of peroxidase and IAA oxidase in hypocotyl of cucumber seedlings was investigated showing a marked inhibition at low brassinosteroid concentrations [175]. Brassinolide (**1**) promoted the activity of ATPase of mung bean seedlings in light [176]. The effect of brassinolide (**1**) and (22*S*,23*S*)-*homo*brassinolide on invertase activity in isolated radish cotyledons was investigated. The observed increase in fresh weight was accompanied by an increase of the activity of the soluble invertase after a lack phase, suggesting the stimulation of the de-novo synthesis of this enzyme [177]. Also effects of brassinosteroids on nitrate reductase activity in rice seedling were reported [178].

The influence of the spectral quality of light on the growth response of intact bean plants to brassinolide (**1**) was investigated, indicating a partially overcome of the natural inhibitory effects of white fluorescent light on stem elongation after brassinosteroid application [179, 180]. Furthermore, experiments with light or in darkness were done to clarify the influence of light on the brassinosteroid induced mung bean epicotyl elongation. Epicotyl growth was inhibited in red, green or white light but promoted in far red light compared with growth in darkness. The growth stimulating effect of brassinosteroid on radish roots or leaves was more enhanced under short-day than under long-day conditions [181]. Meudt et al. [182] found that red light induced plumular hook opening in *Phaseolus vulgaris* was counteracted by brassinosteroids and caused a looping of the hypocotyl hook region. Brassinosteroids were 1000 times more active than IAA and seem to react in a synergistic manner.

Active brassinosteroids also promote the growth of gravisensitive hypocotyls, increase the graviperception of young internode tissues and the bending of gravistimulated sections as well as the subsequent reversal of bending after the sections were placed vertically [183, 184].

Several recent reports indicate clearly, that brassinosteroids change the balance of other endogeneous phytohormones such as IAA, ABA, gibberellins and ethylene [159, 185].

Further publications [179, 186, 187] have established a brassinosteroid induced stimulation of the photosynthetic activity, expressed by enhancement of CO_2-fixation, increasing the protein biosynthesis and the content of reducing sugars. ^{14}C-tracer experiments in rice plants indicated, that the promotion effect of brassinolide (**1**) is connected with a stimulation of translocation of photosynthetic products [188].

First results have been published to clarify the mode of action of brassinosteroids on the molecular level. Thus, Kulaeva et al. [189] published the activation of the protein synthesis in wheat leaves after treatment with (22*S*,23*S*)-*homo*-brassinolide, including the synthesis of specific proteins, some of them correspond to heat shock proteins. Also an increase of thermoresistance of protein synthesis was observed. Very recently, similar induction of specific polypeptides were found in elongating soybean stems and differentiating Jerusalem artichoke [190].

Contrary to the well-documented effects of brassinosteroids on shoot tissue, till now only few investigations have been done with roots showing a quite complex picture, whereas with cuttings an inhibitory action was suggested [141]. With root segments promotion of growth has been reported [168, 191]. In more recent studies, 0,1 mmol 24-*epi*brassinolide (**6**) inhibited adventitious root formation and elongation in both hypocotyl and epicotyl cuttings from mung bean [192] and tomato. In the latter case it was also established that roots are influenced directly and not indirectly via the shoot [193]. Treatment of the scutellum of rice root with 10^{-8} mol/l brassinolide (**1**) stimulated root growth whereas application to the root tip was inhibitory [195]. In soybean hypocotyl segments treatment with 24-*epi*brassinolide (**6**) in the dark caused induction of adventitious roots with a maximum effect at 0,1 ng/g brassinosteroid [195]. In experiments with cloned sprouts of *Matricaria chamomilla*, treatment with 0,01 ppm (22*S*,23*S*)-*homo*-brassinolide stimulated rooting with a 2–2,5 fold increase of root weight [196].

In cell cultivation experiments with *Daucus carota* cultures 24-*epi*brassinolide (**6**) induced cell enlargement without having an effect on cell multiplication [197]. *Brassica napus* seeds, cultured on Murashige-Skoog medium, showed better germination, larger cotyledons and higher root number, but had twisted tops and roots, lower fresh weight and plant height. At 0,1 μg/g the callus formation from cotyledon segments and rooting of the calluses were stimulated, but inhibited by higher concentrations [198]. In the case of *Astragalus adsurgens* hypocotyl, 24-*epi*brassinolide (**6**) induced effectively bud differentiation. The induction rate was higher at 10^{-6} mol/l than at 10^{-4} mol/l [199]. Using callus and suspension cultures of *Agrobacterium tumefaciens* transformed hormone-autonomous *Nicotiana tabacum* the fresh and dry weight and the relative water content of call uses were drastically reduced by application of 24-*epi*brassinolide (**6**) and 22,23,24-tri*epi*brassinolide [200]. This inhibitory effect, although less prominent, has been found also in other tested hormone-auxotroph cell lines, e.g., from wild type tobacco, carrot and Petunia [201].

Gartz et al. [202] reported a strong growth promoting effect of (22*S*,23*S*)-*homo*-brassinolide in mycelial cultures of the Basidiomycete *Psilocybe cubensis*, suggesting that brassinosteroids could also play a physiological role in higher fungi. The enhancement of cyclosporii as well as pigment production of *Tolypocladium inflatum* under the influence of some brassinosteroids was likewise reported [203].

With regard to the structural similarity with ecdysteroids, first investigations concerning a possible physiological effect of brassinosteroids on insects have been carried out. Thus, it was shown that brassinosteroids inhibit the evagination of imaginal discs of the diptera *Phormia terra novae* [204] and exhibit a moult-inhibiting activity after feeding at the last larval instar of the cockroach *Periplaneta americana* [205]. Brassinosteroids, especially (22*S*,23*S*)-*homo*castasterone, bind competitively to an ecdysteroid receptor from larvae of the blowfly *Calliphora vicina* and inhibit biological responses to the moulting hormone 20-hydroxyecdysone [206]. Therefore, they represent the first antiecdysteroids, which could possibly open new pathways to ecologically safe insecticides.

6.4 Practical Aspects

Intensive efforts have been made in the field of practical application of brassinosteroids, most of them are published in the patent literature. Besides brassinolide (**1**) preferably 24-*epi*brassinolide (**6**), 22,23,24-tri*epi*brassinolide and (22*S*,23*S*)-*homo*-brassinolide were investigated as promising candidates. Sometimes, combinations, e.g., with auxin [207, 208], gibberellin [209], salts of choline [208] and calcium cyanamide [210] were applied.

The most investigated effects are growth stimulation, increase of crop yield or other profitable bioregulatory effects on wheat, rice, maize and barley [207, 208, 211–217], Leguminosae including soy bean and azuki bean [208, 211, 215, 218–220], potato [208, 211, 215, 221–223], beet [224], fruits such as citrus [225], peach, persimmon, grape and apple [208], lettuce [217], onion, garlic [208], yam [227], melon [228], egg plant [215] and grass [208, 226]. For Japanese persimmon, it was also found that application of brassinolide (**1**) before flowering reduces the fruit abscission [229].

Field trials and hydroponic studies with tobacco, soaking the seed in 24-*epi*brassinolide (**6**), increased seed oxygen absorption, ethylene production, germination rate, root dry weight and length [230]. For rice, tomato and sweet pepper, after treatment with brassinolide, the freshweight, dry weight and protein content were increased [231]. Comparable effects were determined for beans upon application of brassinolide at the flowering stage [232].

In an extensive field program undertaken in China, 24-*epi*brassinolide (**6**) was sprayed on wheat at its blossoming or filling stage to give a 10% yield increase in more than 100 trials from 1986–1989. With corn, a yield increase of 10–15% was given for most tests [233]. However, a dependence on an average yield per hectar must be considered in all such experiments, and the increase of crop yield seems to be the higher the more non-optimal growth conditions are. Such an antistress hormone functionality of brassinosteroids became more and more interesting for potential practical application [138] and a considerable number of recent papers dealing with this topic appeared. This promising influence of brassinosteroid application is observed not only upon growth under low-yield cultivation conditions, but also involves increased resistance against cold [234] and heat [235], drought, pathogenic infection [236], herbicidal injury [237–239] and salinity [237, 240–242]. Brassinosteroids increased cold resistance of egg plants [138], cucumber and rice [243]. In experiments with navel oranges the abnormal fruit abortion due to a cold environment was reduced [244]. Brassinolide at 0.01 μg/g had a promotive effect on rice plants under environmental stress caused by low temperature and shading conditions [188]. It also promoted the recovery of growth of coleoptyles and mesocotyl of maize after chilling [245] and can partially overcome high temperature-induced suppression of cucumber seed germination [246].

The stimulation of germination of aged rice seeds by pretreatment with brassinolide (**1**) has been reported. The germination rate of 13 year-old rice seeds increased to 84% after treatment compared to 38% in the control or 68% after GA_3 treatment [247]. Retarding effects of 1 ppm 24-*epi*brassinolide (**6**) on maturation and senescence of hypocotyl segments of mung bean seedlings are also reported [248].

The growth of barley embryo under salt stress was stimulated by (22*S*, 23*S*)-*homo*brassinolide and this effect was cumulative when applying a mixture of the brassinosteroid with ethanolamine. The same combination promoted root growing of barley under drought stress [249]. In pot experiments with sugar beet, treatment with (22*S*,23*S*)-*homo*brassinolide compensated mild drought stress completely and the mass of tap-roots was always increased under stress conditions by 25–30% [250].

Tubers of potato artificially infected with *Phytophthora infestans* or *Fusarium sulfuricum* and treated with brassinosteroids showed a higher resistance to phytopathogens than the control and formed more protective substances in response to infection [251].

Of special importance for potential practical application are data about the toxicity of brassinosteroids. For 24-*epi*brassinolide (**6**) the following values detected in the Research Laboratory of Nippon Kayaku Co. were given [252], indicating very favourable toxicological properties:

- acute toxicity oral LD_{50} (mouse) > 1000 mg/kg; (rat) > 2000 mg/kg;
- acute toxicity dermal LD_{50} (rat) > 2000 mg/kg
- fish toxicity TLM48 (carp) > 10 mg/l
- Ames test negative
- a 0.01% solution causes no eye-irritation in rabbit.

7 Conclusions

During the last decade, brassinosteroid research has been an area of tremendous activity reflected in two international symposia held in 1990 at Washington and

Halle/Saale. Applying modern analytical techniques a series of new naturally occurring members including first sugar conjugates were discovered. Their occurrence covers a wide range of plant species including Angiosperms and Gymnosperms, suggesting an ubiquitous distribution. With regard to chemistry, intense efforts were directed to the synthesis of the native brassinosteroids and structural analogs stimulating the development of original approaches for stereoselective construction of such steroidal plant growth regulators. These studies included also pathways to labelled compounds applied for first biosynthetic and metabolic experiments.

Brassinosteroids accelerate metabolic and growth processes at very low concentrations in plants. Their multiple physiological activity involves, among other effects on the cell membrane, production of enzymes and activation of photosynthesis. Brassinosteroids interact with environmental stimuli, e.g., light, gravity and temperature, as well as with the recognized hormones. According to recent results, brassinosteroids are able to trigger the synthesis of particular proteins. Thus, in total, they fulfill the characteristics necessary to constitute a new class of phytohormones [185] although further research on their physiology, function and molecular mechanisms of action is necessary.

Hitherto existing studies on their practical application have shown that brassinosteroid-induced increase of crop yield is obtained preferably under low-yield conditions. Sometimes observed instability in the results during field application is surely due to great gaps in our knowledge in such parameters as formulation, uptake, transport, metabolism and physiology of the applied compound and require further studies. For future application, antistress effects of brassinosteroids seem to be important where promising results in plant cultivation under unfavourable environmental conditions such as dryness, extreme temperature, salinity, pathogenic infection and herbicide treatment are observed. Further potential applications could be expected e.g. improving crop quality, optimizing time of harvest and cultivation area. The high efficiency in very low doses, their non-toxicity as well as their natural character are suitable prerequisites for a further development of brassinosteroids as ecologically soft stress-modulators and bioregulators in agriculture and horticulture.

Acknowledgement: The authors are indebted to Mrs. G. Schmidt and Mrs. M. Süße for accurate drawing of the formulas and Mrs. H. Bottin for typewriting the manuscript.

8 References

1. Grove MD, Spencer GF, Rohwedder WK, Mandava NB, Worley JF, Warthen JD, Steffens GL, Flippen-Anderson JL, Cook JC (1979) Nature (London) 281: 216
2. Mori K (1985) Yuki Gosei Kagaku Kyokaishi 43: 849
3. Adam G, Marquardt V (1987) Z Chem 27: 41
4. Khripach VA (1990) Pure Appl Chem 62: 1319

5. McMorris TC (1989) Analysis of brassinosteroids, antheridiol and oogoniols. In: Nes WD, Parrish EJ (eds) Analysis of sterols and other biologically significant steroids. Academic, San Diego, Calif., p. 61
6. Sembdner G, Schneider G, Schreiber K (eds) (1987) Methoden der Pflanzenhormonanalyse. Springer, Berlin Heidelberg New York
7. Yokota T, Takahashi N (1986) Chemistry, physiology and agricultural application of brassinolide and related steroids. In: Bopp M (ed) Plant growth substances 1985. Springer, Berlin Heidelberg New York, p 129
8. Adam, G, Marquardt V (1986) Phytochemistry 25: 1787
9. Meudt WJ (1987) Chemical and biological aspects of brassinolide. In: Fuller G, Ness WD (eds) Ecology and metabolism of plant lipids. ACS Symposium Series No 325, p 53
10. Mandava NB (1988) Annu Rev Plant Physiol Plant Mol Biol 39: 23
11. Yokota T, Kim SK, Takahashi N (1988) Abstr 13th Int Conf on Plant Growth Substances, Calgary, Abstr 168
12. Kim SK, Akihisa T, Tamura T, Matsumoto T, Yokota T, Takahashi N (1988) Phytochemistry 27: 629
13. Kim SK, Yokota T, Takahashi N (1987) Agric Biol Chem 51: 2303
14. Yokota T, Kim SK, Kosaka Y, Ogino Y, Takahashi N (1987) Conjugation of brassinosteroids. In: Schreiber K, Schuette HR, Sembdner G (eds) Conjugated plant hormones – structure, metabolism and function. Institute of Plant Biochemistry Halle, Academy of Sciences GDR p 288 (Proceedings of the International Symposium Conjugated Plant Hormones, 3–7 Nov 1986 at Gera)
15. Yokota T, Kim SK, Ogino Y, Takahashi N (1987) Proc. 14th Meeting Plant Growth Regulator Soc Amer p 28
16. Sakakibara M, Okada K, Ichikawa Y, Mori K (1982) Heterocycles 17: 301
17. Ikekawa N, Takatsuto S (1984) Shitsuryo Bunseki 32: 55
18. Thompson MJ, Meudt WJ, Mandava NB, Dutky SR, Lusby WR, Spaulding DW (1982) Steroids 39: 89
19. Abe H, Morishita T, Uchiyama M, Takatsuto S, Ikekawa N, Ikeda M, Sassa T, Kitsuwa T, Marumo S (1983) Experientia 39: 351
20. Takatsuto S, Ying B, Morisaki M, Ikekawa N (1981) Chem Pharm Bull 29: 903
21. Yokota T, Baba J, Takahashi N (1982) Tetrahedron Lett 23: 4965
22. Yokota T, Baba J, Koba S, Takahashi N (1984) Agric Biol Chem 48: 2529
23. Mori K, Sakakibara M, Okada K (1984) Tetrahedron 40: 1767
24. Yokota T, Baba J, Takahashi N (1983) Agric Biol Chem 47: 1409
25. Mori K (1985) Rev Latinoam Quim 16: 55
26. Ikekawa N, Nishiyama F, Fujimoto Y (1988) Chem Pharm Bull 36: 405
27. Thompson MJ, Mandava N, Flippen Anderson JL, Worley JF, Dutky SR, Robbins WE, Lusby W (1979) J Org. Chem 44: 5002
28. Yokota T, Arima M, Takahashi N (1982) Tetrahedron Lett 23: 1275
29. Anastasia M, Ciuffreda P, Del Puppo M, Fiecchi A (1983) J Chem Soc, Perkin Trans I 383
30. Yokota T, Kim SK, Fukui Y, Takahashi N, Takeuchi Y, Takematsu T (1987) Phytochemistry 26: 503
31. Takatsuto S, Ikekawa N (1984) J Chem Soc, Perkin Trans I 439
32. Okada K, Mori K (1983) Agric Biol Chem 47: 925
33. Anastasia M, Ciuffreda P, Fiecchi A (1983) J Chem Soc Perkin Trans I 379
34. Mori K, Takeuchi T (1988) Liebigs Ann Chem 815
35. Schneider JA, Yoshihara K, Nakanishi K, Kato N (1983) Tetrahedron Lett 24: 3859
36. Suzuki Y, Yamaguchi I, Yokota T, Takahashi N (1986) Agric Biol Chem 50: 3133
37. Yokota T, Morita M, Takahashi N (1983) Agric Biol Chem 47: 2149
38. Yokota T, Koba S, Kim SK, Takatsuto S, Ikekawa N, Sakakibara M, Okada K, Mori K, Takahashi N (1987) Agric Biol Chem 51: 1625
39. Mori K, Sakakibara M, Ichikawa Y, Ueda H, Okada K, Umemura T, Yabuta G, Kuwahara S, Kondo M, Minobe M, Sogabe A (1982) Tetrahedron 38: 2099
40. Takatsuto S, Ikekawa N (1986) J Chem Soc Perkin Trans I 2269

41. Ikekawa N, Takatsuto S, Kitsuwa T, Saito H, Morishita T, Abe H (1984) J Chromatogr 290: 289
42. Arima M, Yokota T, Takahashi N (1984) Phytochemistry 23: 1587
43. Plattner RD, Taylor SL, Grove MD (1986) J Nat Prod 49: 540
44. Cao H, Chen S, Jing J (1987) Zhiwan Zazho 952
45. Ikekawa N, Takatsuto S, Marumo S, Abe H, Morishita T, Uchiyama M, Ikeda M, Sasa T, Kitsuwa T (1983) Proc Jpn Acad Ser B 59: 9
46. Ikeda M, Takatsuto S, Sassa T, Ikekawa N, Nukina M (1983) Agric Biol Chem 47: 655
47. Park KH, Saimoto H, Nakagawa S, Sakurai A, Yokota T, Takahashi N, Syono K (1989) Agric Biol Chem 53: 805
48. Tada E (1987) JP 1–117 899
49. Yokota T (1987) Shokubutsu no Kagaku Chosetsu 22: 10
50. Baba J, Yokota T, Takahashi N (1983) Agric Biol Chem 47: 659
51. Takatsuto S, Abe H, Gamoh K (1990) Agric Biol Chem 54: 1057
52. Takatsuto S, Omote K, Gamoh K, Ishibashi M (1990) Agric Biol Chem 54: 757
53. Takatsuto S, Yokota T, Omote K, Gamoh K, Takahashi N (1989) Agric Biol Chem 53: 2177
54. Suzuki Y, Yamaguchi I, Takahashi N (1985) Agric Biol Chem 49: 49
55. Yokota T, Arima M, Takahashi N, Crozier A (1985) Phytochemistry 24: 1333
56. Yokota T, Arima M, Takahashi N, Takatsuto S, Ikekawa N, Takematsu T (1983) Agric Biol Chem 47: 2419
57. Schmidt J, Yokota T, Adam G, Takahashi N (1991) Phytochemistry 30: 364
58. Morishita T, Abe H, Uchiyama M, Marumo S, Takatsuto S, Ikekawa N (1983) Phytochemistry 22: 1051
59. Park KH, Yokota T, Sakurai A, Takahashi N (1987) Agric Biol Chem 51: 3081
60. Park KH, Hyun KH (1987) Han'guk Nonghwa Hakhoechi 30: 54
61. Wada K, Marumo S, Abe H, Morishita T, Nakamura K, Uchiyama M, Mori K (1984) Agric Biol Chem 48: 719
62. Takatsuto S, Ying B, Morisaki M, Ikekawa N (1982) J Chromatogr 239: 233
63. Takatsuto S, Ikekawa N (1986) Chem Pharm Bull 34: 3435
64. Abe H, Nakamura K, Morishita T, Uchiyama M, Takatsuto S, Ikekawa N (1984) Agric Biol Chem 48: 1103
65. Schmidt J, Vorbrodt HM, Adam G (1986) Biomed Environ Mass Spectrom 13: 663
66. Gamoh K, Okamoto N, Takatsuto S, Tejima I (1990) Anal Chim Acta 228: 101
67. Gamoh K, Omote K, Okamoto N, Takatsuto S (1989) J Chromatogr 469: 424
68. Gamoh K, Kitsuwa T, Takatsuto S, Fujimoto Y, Ikekawa N (1988) Anal Sci 533
69. Horgen PA, Nakagawa CH, Irvin RT (1984) Can J Biochem Cell Biol 62: 715
70. Schlagnhaufer C, Arteca RN, Phillips AT (1988) Plant Physiol (Suppl) 86: 4
70a. Yokota T, Watanabe S, Ogino Y, Yamaguchi I, Takahashi N (1990) J Plant Growth Regul 9: 151
70b. Zeigan D, Adam G, Porzel A, Kutschabsky L (1990) Abstr Int Workshop Brassinosteroids – Chemistry, Bioactivity, Application, Halle, 29 October–2 November 1990 p 38
71. Kutschabsky L, Adam G, Vorbrodt HM (1990) Z Chem 30: 136
72. Kutschabsky L, Marquardt V, Adam G Z Chem in preparation
73. Ishiguro M, Takatsuto S, Morisaki M, Ikekawa N (1980) J Chem Soc Chem Commun 962
74. Fung S, Siddall JB (1980) J Am Chem Soc 102: 6580
75. Takatsuto S, Ikekawa N (1984) Chem Pharm Bull 32: 2001
76. Mori K (1980) Agric Biol Chem 44: 1211
77. Wada K, Marumo S (1981) Agric Biol Chem 45: 2579
78. Akhrem AA, Lakhvich FA, Khripach VA, Zhabinskii VN, Kovganko NV (1984) Dokl Akad Nauk SSSR 275: 1089
79. Takatsuto S, Kobayashi K, Watanabe T, Kuriyama H, Furuse T (1988) Agric Biol Chem 52: 3217
80. Watanabe T, Kuriyama H, Furuse T, Kobayashi K, Takatsuto S (1988) Agric Biol Chem 52: 2117

81. Aburatani M, Takeuchi T, Mori K (1987) Synthesis 181
82. Aburatani M, Takeuchi T, Mori K (1986) Agric Biol Chem 50: 3043
83. Anastasia M, Allevi P, Ciuffreda P, Oleotti A (1985) Steroids 45: 561
84. Aburatani M, Takeuchi T, Mori K (1987) Agric Biol Chem 51: 1909
85. Barton DHR, Lusinchi X, Magdzinski L, Ramirez JS (1984) J Chem Soc Chem Commun 1236
86. Anastasia M, Allevi P, Ciuffreda P, Fiecchi A, Scala A (1984) J Org Chem 49: 4297
87. Ferrer JC, Lalueza R, Saavedra O, Brosa C (1990) Tetrahedron Lett 31: 3941
88. Anastasia M, Ciuffreda P, Fiecchi A (1982) J Chem Soc Chem Commun 1169
89. Takatsuto S, Yazawa N, Ishiguro M, Morisaki M, Ikekawa N (1984) J Chem Soc Perkin Trans I 139
90. Zhou WS, Jiang B, Pan XF (1988) J Chem Soc Chem Commun 791
91. Zhou WS, Zhou YP, Jiang B (1989) Synthesis 426
92. Brooks CJW, Ekhato IV (1982) J Chem Soc Chem Commun 943
93. Zhou WS (1989) Pure Appl Chem 61: 431
94. Akhrem AA, Lakhvich FA, Khripach VA, Kovganko NV (1983) J Org Chem SSSR 19: 1249
95. Takatsuto S, Ikekawa N (1987) Chem Pharm Bull 35: 829
96. Takatsuto S, Ikekawa N (1987) Chem Pharm Bull 35: 3006
97. Takatsuto S, Muramatsu M, Ohya Y, Hayashi S, Shida A (1988) Agric Biol Chem 52: 2059
98. Cerny V (1989) Collect Czech Chem Commun 54: 2211
99. Edwards JE, Mills JS, Sundeen J, Freed JH (1969) J Am Chem Soc 91: 1248
100. Hutchins RF, Thompson MJ, Svoboda JA (1970) Steroids 15: 113
101. Salmond WG, Sobala MC (1977) Tetrahedron Lett 1695
102. Takatsuto S, Ikekawa N (1983) J Chem Soc, Perkin Trans I 2133
103. Takatsuto S, Yazawa N, Ikekawa N (1984) Phytochemistry 23: 525
104. Takatsuto S, Ikekawa N (1983) Tetrahedron Lett 24: 773
105. Donaubauer JR, Greaves AM, McMorris TC (1984) J Org Chem 49: 2833
106. Hayami H, Sato M, Kanemoto S, Morizawa Y, Oshima K, Nozaki H (1983) J Am Chem Soc 105: 4491
107. Kametani T, Kigawa M, Tsubuki M, Honda T (1988) J Chem Soc Perkin Trans I 1503
108. Kametani T, Katoh T, Tsubuki M, Honda T (1986) J Am Chem Soc 108: 7055
109. Kametani T, Keino K, Kigawa M, Tsubuki M, Honda T (1989) Tetrahedron Lett 30: 3141
110. Zhou WS (1988) Acta Chim Sin 46: 598
111. Zhou WS, Tian WS (1987) Tetrahedron 43: 3705
112. Zhou WS, Jiang B, Pan XF (1989) J Chem Soc Chem Commun 612
112a. Zhou WS, Jiang B, Pan XF (1990) Tetrahedron 46: 3173
113. Yamamoto Y, Nishii S, Maruyama K (1986) J Chem Soc Chem Commun 102
114. Back TG, Brunner K, Krishna MV, Lai EKY (1989) Can J Chem 67: 1032
115. Takahashi T, Ootake A, Yamada H, Tsuji J (1985) Tetrahedron Lett 26: 69
116. Kametani T, Katoh T, Tsubuki M, Honda T (1987) Chem Pharm Bull 35: 2334
117. Kametani T, Katoh T, Fujio J, Nogiwa I, Tsubuki M, Hondo T (1988) J Org Chem 53: 1982
118. Mikami K, Loh TP, Nakai T (1988) J Chem Soc Chem Commun 1430
119. Mikami K, Kawamoto K, Nakai T (1986) Tetrahedron Lett 27: 4899
120. Mikami K, Kawamoto K, Nakai T (1985) Tetrahedron Lett 26: 5799
121. Takatsuto S, Ikekawa N (1986) J Chem Soc, Perkin Trans I 591
122. Takatsuto S, Ikekawa N (1986) Chem Pharm Bull 34: 1415
123. Takatsuto S, Ikekawa N (1986) Chem Pharm Bull 34: 4045
124. Allevi P, Anastasia M, Cerana R, Ciuffreda P, Lado P (1988) Phytochemistry 27: 1309
125. Seo S, Nagasaki T, Katsuyama Y, Matsubara F, Sakata T, Yoshioka M, Makisumi Y (1989) J Labelled Compd Radiopharm 27: 1383
126. Yokota T, Ogino Y, Suzuki H, Takahashi N, Saimoto H, Fujioka S, Sakurai A (1990) Abstr Int Workshop Brassinosteroids – Chemistry, Bioactivity, Application, Halle 29 October–2 November 1990, p 5

127. Kohout L, Velgova H, Strnad M, Kaminek M (1987) Coll Czech Chem Comm 52: 476
128. Kohout L, Strnad M (1989) Collect Czech Chem Commun 54: 1019
129. Kohout L (1989) Coll Czech Chem Commun 54: 3348
130. Adam G, Vorbrodt HM, Marquardt V, Kien NT, Porzel A (1989) Proc 5th Conf Chem Biotechnol Biol Active Nat Prod, Warna 18–23 September 1989, 3: 159
131. Marquardt V, Adam G, Alonso E, Coll F, Creuzburg D (1990) Abstr Int Workshop Brassinosteroids – Chemistry, Bioactivity, Application, Halle 29 October–2 November 1990 p 10
132. Takatsuto S, Ikekawa N (1987) Chem Pharm Bull 35: 986
133. Takatsuto S, Ikekawa N, Morishita T, Abe H (1987) Chem Pharm Bull 35: 211
134. Akhrem AA, Lakhvich FA, Khripach VA, Zhabinskii VN, Kovganko NV (1984) Dokl Akad Nauk SSSR 275: 626
135. Akhrem AA, Lakhvich FA, Khripach VA, Kovganko NV, Zhabinskii VN (1985) Dokl Akad Nauk SSSR 283: 130
135a. Vorbrodt HM, Kien NT, Porzel A, Adam G (1990) Abstr Int Workshop Brassinosteroids – Chemistry, Bioactivity, Application, Halle 29 October–2 November 1990 p 16
136. Hoerhold C, Vorbrodt HM, Boehme H, Porzel A, Adam G (1990) Abstr Int Workshop Brassinosteroids – Chemistry, Bioactivity, Application, Halle 29 October–2 November 1990 p 24
137. Mandava NB, Sasse JM, Yopp JH (1981) Physiol Plant 53: 453
138. Hamada K (1986) FFTC Book Ser 34: 188
139. Strnad M, Kaminek M (1985) Biol Plant 27: 209
140. Mitchell JW, Livingston GA (1968) Agric Handb US Dept Agric 336
141. Yopp JH, Mandava NB, Sasse JM (1981) Physiol Plant 53: 445
142. Hewitt FR, Hough T, O'Neill P, Sasse JM, Williams EG, Rowan KS (1985) Aust J Plant Physiol 12: 201
143. Wada K, Kondo H, Marumo S (1985) Agric Biol Chem 49: 2249
144. Takatsuto S, Yazawa N, Ikekawa N, Takematsu T, Takeuchi Y, Koguchi M (1983) Phytochemistry 22: 2437
145. Strnad M (1990) Abstr Int Workshop Brassinosteroids – Chemistry, Bioactivity, Application, Halle 29 October–2 November 1990, p 37
146. Ikekawa N (1986) Structure and Function of Natural Sterols. In: Imura H, Goto T, Murachi T, Nakajima T (eds) Natural Products and Biological Activities. Univ Tokyo Press, Tokyo p 75
146a. Takatsuto S, Ikekawa N, Morishita T, Abe H (1987) Chem Pharm Bull 35: 211
147. Cerny V, Strnad M, Kaminek M (1986) Collect Czech Chem Commun 51: 687
148. Kakiuchi T, Kamuro Y, Takatsuto S, Kobayashi K (1988) Agric Biol Chem 52: 2381
149. Takatsuto S, Futatsuya F, Kobayashi K, Satoh H (1989) Agric Biol Chem 53: 263
150. Hayashi S, Hohjoh T, Shida A, Ikekawa N (1988) EP O 282 984
151. Arteca RN, Tsai DS, Schlagnhaufer C, Mandava NB (1983) Physiol Plant 59: 539
152. Arteca RN, Schlagnhaufer C (1984) Physiol Plant 62: 445
153. Schlagnhaufer C, Arteca RN, Yopp JH (1984) Physiol Plant 61: 555
154. Schlagnhaufer C, Arteca RN, Yopp JH (1984) Physiol Plant 60: 347
155. Wu YM, Bao YW, Liu Y (1987) Acta Phytophysiologica Sinica 13: 107
156. Arteca RN, Bachman JM, Mandava NB (1988) J Plant Physiol 133: 430
157. Arteca RN, Bachman JM, Tsai DS, Mandava NB (1988) Physiol Plant 74: 631
158. Arteca RN, Bachman JM (1987) J Plant Physiol 129: 13
159. Eun JS, Kuraishi S, Sakurai N (1989) Plant Cell Physiol 30: 807
160. Schlagnhaufer CD, Arteca RN (1985) Physiol Plant 65: 151
161. Schlagnhaufer CD, Arteca RN (1985) Plant Physiol 78: 300
162. Schubert S, Matzke H (1985) Physiol Plant 64: 285
163. Romani G, Marre MT, Bonetti A, Cerana R, Lado P, Marre E (1983) Physiol Plant 59: 528
164. Marre E (1985) Fusicoccin-induced and hormone-induced changes of hydrogen extrusion physiological implications. In: St. John JB, Berlin E and Jackson PC (eds) Beltsville Symposia in Agricultural Research, 9. Frontiers of Membrane Research in Agriculture; Beltsville, MD, USA, 20–24 May 1984 p 439

165. Cerana R, Colombo R, Lado P (1983) Physiol Veg 21: 875
166. Cerana R, Bonetti A, Marre MT, Romani G, Lado P, Marre E (1983) Physiol Plant 59: 23
167. Cerana R, Lado P, Anastasia M, Ciuffreda P, Allevi P (1984) Z Pflanzenphysiol 114: 221
168. Cerana R, Spelta M, Bonetti A, Lado P (1985) Plant Sci 38: 99
169. De Michelis MI, Lado P (1986) Physiol Plant 68: 603
170. Dahse I, Sack H, Bernstein M, Petzold U, Mueller E, Vorbrodt HM, Adam G (1990) Plant Physiol 93: 1268
171. Dahse I, Willmer CM, Grammatikopoulus G, Nikolopoulos D (1990) Abstr Int Workshop Brassinosteroids – Chemistry, Bioactivity, Application, Halle 29 October–2 November 1990 p 34
172. Kalinich JF, Mandava NB, Todhunter JA (1985) J Plant Physiol 120: 207
173. Kalinich JF, Mandava NB, Todhunter JA (1986) J Plant Physiol 125: 345
174. Mandava NB, Thompson MJ, Yopp JH (1987) J Plant Physiol 128: 53
175. Xu RJ, Zhao YJ (1989) Acta Phytophysiologica Sinica 15: 263
176. Henry EW, Dungy LJ, Bracciano DM (1981) Proc Plant Growth Regul Soc Amer p 146
177. Beinhauer K, Andreas W, Creuzburg D, Lang S, Vorbrodt HM, Adam G (1990) Abstr Int Workshop Brassinosteroids – Chemistry, Bioactivity, Application, Halle 29 October–2 November 1990 p 22
178. Mai Y, Lin J, Zeng X, Pan R (1989) Zhiwu Shenglixue Tongxun 50
179. Krizek DT, Mandava NB (1983) Physiol Plant 57: 324
180. Krizek DT, Mandava NB (1983) Physiol Plant 57: 317
181. Kamuro Y, Inada K (1987) Proc Ann Plant Growth Regul Soc Amer 1987 p 221
182. Meudt WJ, Porath D (1986) Proc Plant Growth Regul Soc Amer p 129
183. Meudt WJ (1986) Plant Physiol (Suppl 4) 80: 9
184. Meudt WJ (1987) Plant Physiol 83: 195
185. Sasse JM (1990) Abstr 200th ACS National Meeting, Symposium on Brassinosteroids, Washington 26–31 Aug 1990 Abst No 119
186. Braun P, Wild A (1984) J Plant Physiol 116: 189
187. Braun P, Wild A (1984), Proc Int Congr Photosynth, 6th, Nijhoff, The Hague, Neth p 461
188. Fujii S, Hirai K, Saka M (1990) Abstr 200th ACS National Meeting, Symposium on Brassinosteroids, Washington 26–31 Aug 1990 Abstr No 79
189. Kulaeva ON, Burkhanova EA, Fedina AB, Danilowa NV, Adam G, Vorbrodt HM, Khripach VA (1989) Dokl Akad Nauk SSSR 305: 1277
190. Clouse S, Zurek D, Hall A (1990) Abstr 200th ACS National Meeting, Symposium on Brassinosteroids, Washington 26–31 Aug 1990, Abstr No 139
191. Roddick JG, Guan M (1990) Abstr Int Workshop Brassinosteroids – Chemistry, Bioactivity, Application, Halle 29 October–2 November 1990 p 19
192. Guan M, Roddick JG (1988) Physiol Plant 73: 426
193. Guan M, Roddick JG (1988) Physiol Plant 74: 720
194. Radi SH, Maeda E (1988) Japanese Journal of Crop Science 57: 191
195. Sathiyamoorthy P, Nakamura S (1990) Plant Growth Regul 9: 73
196. Vorbrodt HM, Donath P, Adam G, Luckner M, Neumann D, Beinhauer K (1988) Ger (East) DD 271 445
197. Bellincampi D, Morpurgo G (1988) Plant Science, 54: 153
198. Chang JQ, Cai DT (1988) Oil Crops of China 18
199. Chen JC, Chi JF (1986) Plant Physiol Commun 51
200. Roth PS, Bach TJ, Thompson MJ (1989) Plant Science (Shannon) 59: 63
201. Bach TJ, Roth PS, Thompson MJ (1990) Abstr Int Workshop Brassinosteroids – Chemistry, Bioactivity, Application, Halle 29 October–2 November 1990 p 47
202. Gartz J, Adam G, Vorbrodt HM (1990) Naturwissenschaften 77: 388
203. Graefe U (1990) Abstr Int Workshop Brassinosteroids – Chemistry, Bioactivity, Application, Halle 29 October–2 November 1990 p 23
204. Hetru C, Roussel JP, Mori K, Nakatani Y (1986) C R Acad Sci, Ser 2 302: 417
205. Richter K (1990) Abstr Int Workshop Brassinosteroids – Chemistry, Bioactivity, Application, Halle 29 October–2 November 1990 p 51

206. Lehmann M, Vorbrodt HM, Adam G, Koolman J (1988) Experientia 44: 355
207. Genma T (1989) JP 1-106802
208. Kajita F, Furushima M, Takematsu T (1985) JP 62-77305 (1986) EP 0220514
209. Kaneda T et al (1989) JP 1-256330
210. Kuraishi S, Eun JS, Sakurai N, Kojima K, Sugiyama K, Sumi T (1990) Abstr Int Workshop Brassinosteroids – Chemistry, Bioactivity, Application, Halle 29 October–2 November 1990 p 42
211. Kumura A, Ishii R, Luo BS, Adachi M, Hamada K, Fujita F (1985) JP 60-116537
212. Kumura A et al (1986) JP 61-76402
213. Hirai Y, Fujii S, Kamuro Y (1985) JP 62-63502
214. Hirai Y, Fujii S (1985) JP 62-67005
215. Mori K, Kishi M (1985) JP 62-135459
216. Takematsu T, Ikekawa N, Shida A (1986) EP 218945 (1988) JP 63-54302
217. Meudt WJ, Thompson MJ, Mandava NB, Worley JF (1984) Can 1173659
218. Kumura A et al (1986) JP 61-277601
219. Genma T (1989) JP 1-146805
220. Hayashi S, Hohjoh T, Furuse T, Kuriyama H, Watanabe T, Takatsuto S (1988) EP 261656
221. Kumura et al (1987) JP 62-67006
222. Genma T (1988) JP 63-170306; WO 88/04890
223. Adam G, Bergmann H, Lang S, Meisgeier G, Vorbrodt HM (1989) Ger (East) DD 265316
224. Genma T (1989) JP 1-146806
225. Takahara T, Hirose K, Ono Y, Iwagaki I, Hirai Y (1985) JP 62-63501
226. Genma T (1989) JP 1-163105
227. Genma T (1989) JP 1-146804
228. Ikekawa N (1988) JP 63-243002
229. Maotani T, Suzuki A, Nishimura T, Kumomato O, Oshima K, Yamanaka Y (1989) J Jpn Soc Hortic Sci 58: 557
230. Han JF, Qi QG, Zhang XM, Zhao YY (1988) Proc 9th Intern Tobacco Scientific Congress 9–13 October 1988, Guangzhou, China p 197
231. Lim UK (1985) Proc Plant Growth Regul Soc Amer p 213
232. Nakaseko K, Yoshida K (1989) Mem Fac Agric Hokkaido Univ 16: 347
233. Zhao YJ, Ikekawa N (1990) Abstr Int Workshop Brassinosteroids – Chemistry, Bioactivity, Application, Halle 29 October–2 November 1990 p 27
234. Hirai Y, Fujii S, Honjo K (1988) JP 63-135303
235. Ikekawa N (1988) JP 63-239201
236. Adachi A, Yamada Y, Fujita F, Nakazawa Y, Amano T, Watanabe K (1984) JP 61-76404
237. Takematsu T (1988) JP 63-66104
238. Takematsu T (1988) JP 63-130504
239. Adachi A, Hamada K, Nishi S, Fujita F, Uesono T, Fujiwara S (1984) JP 61-76403
240. Uesono T, Adachi A, Hamada K, Nishi S, Fujita F, Fujiwara S (1985) JP 62-48602
241. Khokhlova VA, Bokebayeva GA, Adam G, Kulaeva ON (1990) Abstr Int Workshop Brassinosteroids – Chemistry, Bioactivity, Application, Halle 29 October–2 November 1990 p 44
242. Bokebayeva GA, Khokhlova VA, Khripach VA, Adam G, Kulaeva ON (1990) Abstr Int Workshop Brassinosteroids – Chemistry, Bioactivity, Application, Halle 29 October–2 November 1990 p 45
243. Hirai K, Fujii S, Kamuro Y (1984) Abstr Ann Meeting Chem Regul of Plants (Tokyo) p 36
244. Sugiyama K, Kuraishi S (1989) Acta Horticulturae 345
245. He RY, Wang XS, Wang GJ Abstr 200 ACS National Meeting, Symposium on Brassinosteroids, Washington 26–31 Aug 1990 Abstr No 100
246. Katsumi M (1990) Abstr Int Workshop Brassinosteroids – Chemistry, Bioactivity, Application, Halle 29 October–2 November 1990 p 29

247. Yamaguchi T, Wakizuka T, Hirai K, Fujii S, Fujita A (1987) Proc Plant Growth Regul Soc Amer p 26
248. Zhao YJ, Luo WH, Wang YQ, Xu RJ (1987) Acta Phytophysiologica Sinica 13: 129
249. Bergmann H, Meisgeier G, Adam G, Vorbrodt HM, Lang S, Wozniak H Ger (East) DD 277828
250. Schilling G, Schiller C (1990) Abstr Int Workshop Brassinosteroids – Chemistry, Bioactivity, Application, Halle 29 October–2 November 1990 p 20
251. Korableva N, Sukhova L, Dogonadse (1990) Abstr Int Workshop Brassinosteroids – Chemistry, Bioactivity, Application, Halle 29 October–2 November 1990 p 32
252. Ikekawa N (1990) Abstr Int Workshop Brassinosteroids – Chemistry, Bioactivity, Application, Halle 29 October–2 November 1990 p 1

Biochemical Basis of Herbicide Resistance

Kevin C. Vaughn and Stephen O. Duke
United States Dept. of Agriculture, Agric. Research Service,
Southern Weed Science Laboratory, Stoneville, Mississippi 38776, U.S.A.

Chemistry of Plant Protection 7

1 Introduction

Interest in the subject of herbicide resistance has blossomed since the discovery of the first triazine-resistant weed in the early 1970s. Two of the more important reasons for this increased interest are the large numbers of naturally-occuring resistant biotypes (from one in the early 1970s to over 250 biotypes in 1989 [1]) and the large number of herbicide-resistants mutants that have been selected in cell culture, by mutation of the "plant-geneticists' fruit fly" *Arabidopsis thaliana*, and by classical plant breeding protocols [2]. The bourgeoning of biotechnology in the plant sciences also has kindled interest in herbicide resistance. The resistance trait represents an excellent marker for transformation of crop plants because the herbicide may be utilized as a selective agent to remove non-transformed individuals. Additionally, crop plants with herbicide resistance are being pursued by both seed companies and the chemical industry. Recent symposia on the subject of herbicide resistance of the American Chemical Society (1988), the Weed Science Society of America (1989), the Long Ashton Meetings (1989), and the Australian Weed Conference (1990) have been very well attended and give further proof of the interest in this area.

In this review, we will describe the results of studies of both naturally-occurring and laboratory-select ions of herbicide-resistant plant species. For many of these resistant mutants, the mechanism(s) by which resistance is conferred is unknown. Clearly, this will be an area of high research effort for many years.

2 Amino Acid Synthesis Inhibitors

Only three enzymes involved in amino acid synthesis are target sites of commercial herbicides. Glyphosate inhibits 5-enolpyruvylshikimate-3-phosphate synthase (EPSPS), an enzyme of the shikimic acid pathway involved in synthesis of aromatic amino acids (tyrosine, phenylalanine, and tryptophan). The sulfonylureas and imidazolinones inhibit acetolactate synthase (ALS; also called acetohydroxy acid synthase – AHAS), and enzyme required for synthesis of the branched chain amino acids (leucine, isoleucine, and valine). Glufosinate inhibits glutamine synthetase (GS), the final enzyme in the assimilation of inorganic to organic nitrogen. These sites are shown in relationship to the primary metabolism of the cell in Fig. 1.

All three enzymes function primarily in the plastid (chloroplasts, amyloplasts, leucoplasts, or proplastids), although these enzymes are all nuclear-encoded. In each of these cases, there is no alternative metabolic route for synthesis of the amino acid product(s). There is wide latitude in the propensity of higher plants to become resistant to herbicides affecting each of these enzymic sites. Each will be considered separately.

2.1 Glyphosate – Aromatic Amino Acids

In addition to the three essential aromatic amino acids, phenylalanine, tyrosine, and tryptophan, the shikimate pathway produces many aromatic secondary plant

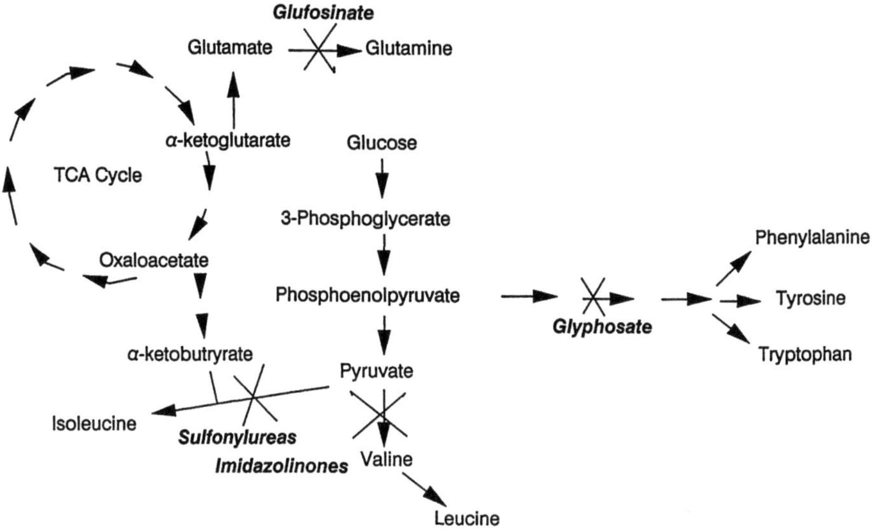

Fig. 1. Sites of action of amino acid synthesis-inhibiting herbicides

products (e.g. lignins, alkaloids, flavonoids, benzoic acids, etc.) that are important in both plant growth and development and in interactions with other organisms. As much as 20% of the carbon fixed by photosynthetic plants flows through this highly-regulated pathway [3]. All of the enzymes of the pathway have been found in the plastid; however, cytosolic forms of some of the enzymes apparently exist. The metabolic role of the enzymes of the shikimate pathway found in the cytosol is not understood, although evidence for dual (cytosolic and plastidic) shikimate pathways exists [4]. The plastidic enzymes are highly regulated, whereas those of the cytosol are under less stringent control. Synthesis of aromatic amino acids is strongly enhanced by light, even in non-photosynthetic tissues [5, 6].

Glyphosate [*N*-(phosphonomethyl)glycine] is the only compound developed as a herbicide which blocks an enzymatic step of the shikimate pathway. Its effects on plants and mechanism of action have been reviewed from various perspectives [7–9]. Physiochemical properties of the glyphosate anion, the active form of the molecule, make it highly mobile in the phloem. It is readily translocated from foliar sites of application with photosynthate to distant metabolic sinks. This, coupled with its relatively slow activity, make it an ideal herbicide for control of perennial weeds with subterranean regenerative organs such as rhizomes or tubers.

This non-selective herbicide apparently has only one significant site of action. It inhibits EPSPS, the enzyme that condenses shikimate-3-phosphate [S3P] and phosphoenolpyruvate [PEP] to yield EPSP and inorganic phosphate [8]. Inhibition is competitive with respect to PEP with a K_i of 0.08 to 0.50 mM and uncompetitive with respect to S3P with a K_i of 10 mM [7, 10]. The inhibition of EPSPS by glyphosate is complex, but accumulated data indicate that glyphosate binds to the region of EPSPS that binds the phosphate moiety of PEP [7]. EPSPS of glyphosate-resistant petunia mutants have little change in their K_m values for EPSPS or S3P; however, the K_m for PEP of the glyphosate-resistant EPSPS is more than 40-fold higher than that of the wild type [7]. Furthermore, production of a thiocyano-EPSPS does not alter the K_m of S3P, whereas its I_{50} for glyphosate and K_m for PEP are both increased [11]. Glyphosate binds the EPSPS-S3P complex 115-fold tighter and 20-fold slower than PEP, and the dissociation rate is 2300-fold slower than that of PEP [12]. Recent data indicate that glyphosate binds the active site of the enzyme with S3P to mimic the tetrahedral intermediate of the enzyme reaction [13]. Through chemical modification investigations, Lys, Arg, Cys, Glu, and His residues are known to be components of the binding site(s) of EPSPS [7, 11, 14, 15], and that an Arg residue (either Arg28 and/or Arg131), the Glu418 residue, and the Cys408 residue are components of the glyphosate-binding site [11, 14, 15]. By site-directed mutagenesis, modification of the Lys23 to Ala or Glu inactivated the petunia EPSPS [16]. Replacement of the Lys23 with Arg produced an enzyme with normal K_m values for its substrates, but with an I_{50} for glyphosate that was 5-fold higher than the unmodified enzyme. An amino acid change of Pro to Ser in the 101 position of EPSPS from *Salmonella typhimurium* confers glyphosate resistance [17].

In green plants, EPSPS is a nuclear-encoded protein with a transit peptide component that allows movement into the plastid. For instance, in *Petunia hybrida*, EPSPS is synthesized with a 72 amino acid transit peptide attached to the 444 amino acid mature enzyme [18]. The molecular mass of the transit peptide for *Corydalis sempervirens* is about 8400 M_r [19]. The nucleotide sequences of several genes coding for EPSPS have been determined, and there is a high level of conservation of amino acid sequences in the mature enzyme. The transit peptides, however, are highly divergent [7, 20]. EPSPS is the only nuclear-encoded, plastid enzyme known to have activity in the cytoplasmic precursor state (pEPSPS) before it is transported into the plastid [21].

Herbicide binding and inhibition can occur in the cytoplasm. Binding of glyphosate to pEPSPS has been shown to inhibit plastid uptake and processing of EPSPS [22]. Import into the plastid is completely inhibited by 10 mM glyphosate. The effect is apparently a direct consequence of glyphosate binding to the pEPSPS-S3P complex, because glyphosate was without effect in the absence of S3P, and glyphosate had no effect on the transport and processing of other nuclear-encoded proteins. The structural stability of the pEPSPS-S3P complex with glyphosate precludes the required conformational flexibility for transport across the plastid envelope.

Although glyphosate has been used extensively and continuously in many areas for over 15 years, no herbicide-resistant weeds have been identified. There are several possible reasons for this. First, there is no evidence that any higher plants enzymatically degrade this herbicide [8]. Apparently higher plants do not have the C−P lyase with which microbes degrade glyphosate. Therefore, there is no degradative enzyme for which amplification or higher efficiency can be selected. Second, glyphosate-resistant EPSPS is much less efficient than susceptible EPSPS. Thus, in order for the mutant form to be adequate for the plant's metabolic needs, another mutation must cause over-expression or gene amplification in order to provide sufficient EPSPS activity. A requirement for a double mutation greatly decreases the odds of the genesis of a resistant weed. Third, glyphosate is tightly bound to soil and is thereby inactivated. Thus, the selection pressure is short-lived, compared to soil-active herbicides with half-lives of weeks. Lastly, if glyphosate does have another site of action which requires higher concentrations of the herbicide than are needed to affect EPSPS, a plant unaffected by glyphosate at the EPSPS site might still be affected in parts of the plant where glyphosate concentrations are high, putting the affected weed at a competitive disadvantage.

Glyphosate is an ideal herbicide for development of herbicide-resistant crops because herbicide-resistant weeds are unlikely to arise in such crops for reasons given above. Also, it is a very toxicologically and environmentally safe herbicide [8]. Genetic engineering can circumvent some of the obstacles to the development of resistance. Three strategies for developing a glyphosate resistant crop species are possible: a) overproduction of EPSPS so that higher concentrations of the herbicide are needed to saturate all sites, b) introduction or induction of an amplified glyphosate-resistant in the plant, or c) introduction of a microbial enzyme that metabolizes glyphosate. The first two of these strategies have been the focus of successful efforts, while the latter approach has, thus far, been unsuccessful.

Selection of glyphosate-tolerant plant cell lines has generally led to selection for mutants that overproduce EPSPS. Elevated EPSPS activity has been found in plant cell cultures of tobacco [23], carrot [24], tomato [25], *Corydalis sempervirens* [26], and petunia [27] selected by culturing surviving plant cells on media of increasing glyphosate concentrations. Plants regenerated from these cultures generally retain some degree of glyphosate tolerance [18, 24–26]. In several glyphosate-tolerant cell lines of tobacco, tolerance to the herbicide was correlated with specific activity of EPSPS extracted from untreated cells (Fig. 2). This was

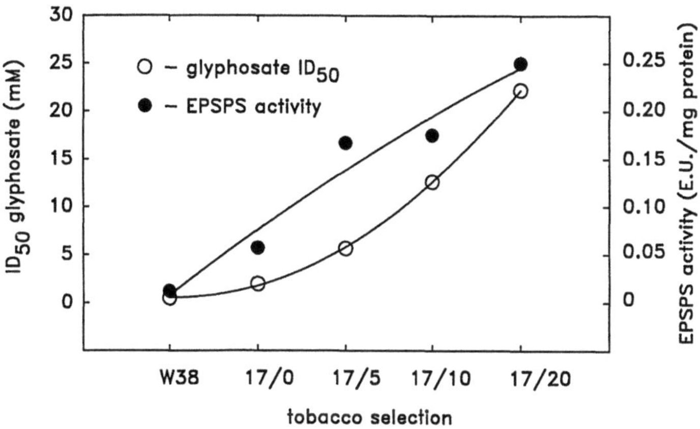

Fig. 2. Effects of glyphosate on growth (ID_{50} = level needed to inhibit growth by 50%) and extractable EPSPS activity from wild type (W38) and four glyphosate-tolerant selections of tobacco cell cultures. Adapted from ref. 23

due to increased content of EPSPS protein and mRNA in these cells [23]. The elevated EPSPS activity was the result of overproduction of the enzyme in the two cases that have been examined at this level [18, 27]. In petunia and tobacco, overproduction was due to gene amplification (i.e., increased copies of the EPSPS gene) [23, 27]. This was determined by analysis of the genome of the tolerant lines using genomic blot analysis with EPSPS cDNA. In *C. sempervirens*, overproduction was not due to gene amplification [28]. In this case, a ten-fold increase in EPSPS mRNA was found in a glyphosate tolerant *C. sempervirens* cell culture that had a 30-fold increase in extractable EPSPS activity. However, Southern and dot blot analyses of the glyphosate-tolerant and -susceptible cell lines revealed that gene amplification was not involved in the over-expression of EPSPS in selected cells. The increased levels of EPSPS mRNA could be due to increased stability and/or a more active promoter region regulating transcription of the EPSPS genes in tolerant cell lines. Overproduced EPSPS was found by immunocytochemical methods, at the transmission electron microscope level, to reside entirely in the plastid [29]. However, the finding that most of the EPSPS in cell lines overproducing it is present as pEPSPS [19] is difficult to reconcile with this result, unless the primary antigenic site for the antibody used is present on EPSPS but not on present on pEPSPS. Also, their study [29] only compared immunogold staining density in plastids versus cytoplasm without consideration of the relative volumes of the plastids versus cytoplasm.

The overproduction strategy has been deliberately implemented by genetically-engineered amplification of the gene encoding EPSPS. Initially, *E. coli* was made eight-fold more tolerant to glyphosate by introduction of a gene encoding EPSPS (the *aro*A gene) on a high-copy plasmid [30]. Later, petunia callus transformed with EPSPS, coupled with the highly active cauliflower mosaic virus promoter, produced about 20-fold more EPSPS than non-transformed plants [18]. The transformed callus was highly tolerant to glyphosate. Petunia plants regenerated from these calli were tolerant to concentrations of commercially formulated glyphosate about 4-fold the minimum lethal dose to plants regenerated from non-transformed callus. Similar methods have produced similar results with *Arabidopsis thaliana* [31] and flax *(Linum usitatissimum)* [32].

Unfortunately, overproduction of the site of action is not a good strategy for a highly phloem-mobile, metabolically stable herbicide like glyphosate. The concentrations of the herbicide in metabolic sinks such as meristematic regions can overwhelm the overexpressed enzyme, even though other parts of the plant are unaffected by the herbicide. For this reason, these plants mature later in the presence of glyphosate than untreated tolerant plants [7]. Other phytotoxicity problems could be expected to be found in flowers and fruits for the same reason. However, the greater amounts of EPSPS in flowers than in leaves [20] in some species could counteract the higher accumulations of glyphosate in these tissues.

The second strategy for producing glyphosate-resistant plants, the introduction of glyphosate-resistant EPSPS, avoids the problems described above. Glyphosate-resistant EPSPS has been found in *Salmonella typhimurium* [17], *Aerobacter aerogenes* [33], and *E. coli* [34]. All of these mutant EPSPSs have higher

K_m values for PEP than the corresponding wild-type enzymes, and the I_{50} values for glyphosate have ranged from 9- to almost 8000-fold higher for the mutant enzyme than the wild types. Transgenic tobacco plants were regenerated from cells transformed with the gene for glyphosate-resistant EPSPS from *E. coli* (with a cauliflower mosaic virus promoter) alone or fused with the cDNA for the transit peptide of petunia EPSPS [35]. The enzyme was found to be entirely cytoplasmic without the transit peptide, and to be highly expressed in the chloroplast with the transit peptide. Tobacco plants transformed with the latter construct were more resistant to glyphosate than those without the capability of transporting resistent EPSPS into the plastid [36]. A chimeric gene of glyphosate-resistant *S. typhimurium* EPSPS (mutant *aro*A gene) and a promoter sequence was used to transform tobacco [37], tomato [38], and *Populus* [39, 40] cells. The plants generated from these transformed cells were only moderately tolerant to glyphosate, perhaps because, without a transit peptide, the enzyme was entirely cytoplasmic, rather than plastidic as it should be for proper functioning in a higher plant cell [41]. The transit peptide of the small subunit of RuBP carboxylase was linked to the glyphosate-resistant *S. typhimurium* EPSPS; however, it did not affect the import of the procaryotic version of EPSPS into tobacco chloroplasts [42]. The transit peptide plus 24 amino acids of the small subunit of RuBP carboxylase fused to EPSPS were transported into the chloroplast both in vivo and in vitro. Plants regenerated from such transformed cells are more resistant to glyphosate than those produced earlier containing only cytoplasmic expression of the *aro*A gene [43].

The gene for EPSPS is highly regulated in a tissue specific manner [20, 44]. Histochemical staining of petunia tissues transformed with the promotor region of EPSPS of petunia fused with the β-glucuronidase (GUS) reporter gene, revealed intense staining in the upper and lower leaf epidermis, vascular tissue, and the mesoderm of flower petals – all known sites of highly active secondary metabolism. Fusion of the promoter region of EPSPS from higher plants with a microbial gene for resistant EPSPS could provide the highly regulated level of gene expression required for normal plant development.

Glyphosate-resistant higher plant EPSPS has been generated by site-directed mutagenesis [7]. The K_m for PEP was about 40-fold higher and the K_i for glyphosate was almost 4 orders of magnitude higher for the EPSPS of glyphosate-resistant than susceptible petunia. Because the K_m for PEP was increased dramatically, one would expect that this enzyme would have to be over-expressed in order for the plant to have normal shikimate metabolism. This is the only reported glyphosate-resistant EPSPS from a higher plant.

The third strategy for producing glyphosate-resistant plants, introducing genes for enzymes that metabolize glyphosate, could overcome the problem of glyphosate accumulation in harvested portions of the crop. The advantages of this strategy are several. First, the herbicide will not accumulate in edible metabolic sinks (fruits, tubers, etc.). Second, the introduced enzyme does not have to be highly regulated in order to function properly. Finally, transformation with only one gene should be required. This strategy has not been successful yet, although the genes for degradation enzymes are available in a variety of microorganisms (e.g. 45, 46). The introduced C–P lyase might degrade other compounds and may be detrimental to the plant unless it is glyphosate-specific.

2.2 Sulfonylureas and Imidazolinones – Branched Chain Amino Acids

Herbicides that inhibit ALS (the sulfonylureas) have been commercially available since 1982 [47]. They were soon followed by the imidazolinones. At least one other herbicide group that inhibits ALS (the 1,2,4-triazolo[1,5A]pryimidines or sulphonanilides) is under development. Molecular characterization of resistance to these herbicides has been previously reviewed [48]. Resistance to imidazolinones and sulfonylureas appeared in some areas within three to five years after use began.

ALS, the first enzyme of the branched chain amino acid pathway, condenses two pyruvate molecules to form CO_2 and 2-acetolactate, the precursor for leucine and valine. The enzyme can also perform the same reaction with pyruvate and 2-ketobutryate to form CO_2 and 2-acetohydroxybutyrate, a precursor of isoleucine. ALS activity requires FAD, thiaminepyrophosphate (TPP), and Mg^{2+} or Mn^{2+} and generates hydroxyethyl-TPP (HETPP) as an intermediate [7]. In the presence of FAD, ALS is predominantly in a tetrameric form [49]. It is regulated through feedback inhibition by all three of its amino acid products [50, 51]. Two forms of ALS exist in some, if not all, higher plants [52–55]. Each form has a characteristic sensitivity to branched chain amino acids and to ALS-inhibiting herbicides. The entire pathway can be found within the plastid, and, as with the shikimate pathway, the enzymes are nuclear encoded, and have transit peptides for proper movement into the plastid. Genes encoding ALS have been isolated and characterized from tobacco and *Arabidopsis* [56], with about 85% homology between the 667 and 670 amino acid proteins, respectively.

Pyruvate oxidase and ALS have a high level of amino acid sequence homology and it has been suggested that the quinone binding site of pyruvate oxidase and the sulfonylurea binding site of ALS may have a common evolutionary origin [7, 57]. Ubiquinone homologues of the native pyruvate oxidase ubiquinone effectively inhibit ALS and will compete with radiolabelled sulfometuron-methyl for this site on the ALS molecule [57]. The structurally diverse array of ALS inhibitors with apparently overlapping binding sites at a quinone-binding site is similar to the situation with the quinone (in this case plastoquinone) binding site of the D-1 protein of photosystem II.

Each herbicide class has a unique set of interactions with the enzyme. All of these compounds can be considered growth inhibitors that cause plant death within a period ranging from several days to more than a week. Most of these compounds are readily absorbed and translocated. Generally, the effects of these compounds are faster than those of glyphosate.

Most of the research on the biochemical interaction of these herbicides with ALS has been done with microbial forms of the enzyme because the plant enzyme is much less stable. However, higher plant ALSs are generally about an order of magnitude more sensitive to both sulfonylureas and imidazolinones than are bacterial ALSs [47]. All of these compounds exhibit binding that is termed "slow-tight binding" [58]. That is, the initial K_i is much higher than that of the terminal K_i. For example, the initial K_i of sulfometuron methyl for *Salmonella typhimurium* ALS is 660 nM, but it is eventually reduced to 65 nM [59]. Similarly, inhibition of pea ALS by sulfonylureas is biphasic with inhibition increasing with time [60]. In vivo, inhibition eventually becomes irreversible and the "slow-tight binding" has been hypothesized to be due to slow inactivation of ALS [91]. The imidazolinones are much weaker ALS inhibitors in vitro than the sulfonylureas, with I_{50}s that are generally more than two orders of magnitude higher than those for sulfonylureas [47]. When the enzyme is incubated with the herbicide in vitro for an extended time, activity is not recovered upon desalting, indicating that the herbicide does not readily dissociate from the enzyme [61].

The sulfonylureas competitively bind the ALS-FAD-TPP-Mg^{2+}-decarboxylated pyruvate complex at a site that overlaps the binding site of the second pyruvate [62]. Inhibition of imidazolinones has been reported to be noncompetitive with respect to pyruvate with corn [63] and pea ALS [64], in contrast to the competitive inhibition of the sulfonylureas.

Tolerance of the crop species for which particular sulfonylureas and imidazolinones are designed is based on rapid metabolism of the herbicide [47].

Sulfonylurea-resistant forms of plant ALS have been produced by mutagenesis and cell culture selection methods with tobacco [60], flax [65], the cyanobacterium *Synechococcus* [66], *Chlamydomonas reinhardtii* [67, 68], and by microspore mutagenesis methods with *Brassica napus* L. [69]. The selected tobacco cell lines were more than 100-fold more resistant to chlorsulfuron than the wild type, although the ALS was 300 times more resistant. This selection was for a single gene mutant in which only one of the ALS forms was resistant to the herbicide. The homozygous mutants were highly resistant, whereas heterozygous genotypes had an intermediate level of resistance [54]. Tobacco is resistant if either of its two ALS forms are resistant [53, 55, 70]. In homozygous mutants for either ALS form, only about one-half of the extractable ALS is inhibited by sulfonylurea herbicides. The two nuclear genes conferring resistance in tobacco are unlinked and both are semidominant [54]. The *SURB-Hra* gene of tobacco isolated by a series of two genetic selections, contains amino acid substitutions (Pro196 to Ala and Trp573 to Leu), whereas the *SURA-C3* has only one substitution (Pro196 to Gln) [2, 70].

Similar results have been obtained with whole plant mutagenesis and selection with *Arabidopsis* [71]. In this case, resistance was due to a single dominant nuclear gene. The gene encoding sulfonylurea-resistant ALS was cloned from the plants and found to differ by only one base pair (resulting in conversion of Pro197 to Ser) from the susceptible plants [72]. When introduced into susceptible tobacco, the gene for sulfonylurea-resistant ALS from *Arabidopsis* provided tobacco with a high level of resistance to sulfonylurea herbicides at the whole plant level [73]. Mutagenesis of seeds and whole plant selection in soybean has led to sulfonylurea herbicide tolerance that is apparently based on enhanced herbicide detoxication

[74] and to a resistant line with ALS that is insensitive to sulfonylureas [75]. In the latter case, resistance is monogenic and semidominant.

Mutants tolerant to imidazolinones have been selected from cultured corn cells [76] and by microspore mutagenesis and selection with canola [77]. The imidazolinone resistance in corn is due to a resistant ALS and these plants are also resistant to sulfonylureas. Development of the corn germplasm into commercial lines is underway. The molecular basis for the imidazolinone-resistant canola has not been reported, however, the genetics are similar to tobacco sulfonylurea resistance – two unlinked, semidominant genes, that confer maximal resistance when a hybrid with both mutations is produced. No yield or quality penalty has been found with these canola mutants.

Examination of cross resistance between sulfonylurea-, triazolopyrimidine-, and imidazolinone-resistant plants reveals no consistent pattern; i.e. some sulfonylurea-resistant plants or plant cell cultures are cross resistant to imidazolinones, some are not, and vice versa [48, 49, 67, 78–81]. *Chlamydomonas* lines resistant to imidazolinones are reportedly even more sensitive to sulfonylureas than the wild type strain from which it was derived [67]. Conversely, a sulfonylurea-resistant culture of *Datura innoxia* was produced that is more sensitive to imidazolinones than the wild type [80]. Furthermore, the differences in K_is of sulfonylureas and imidazolinones are about two orders of magnitude, whereas, in competition experiments using radiolabeled sulfometuron, binding to ALS differs by about three orders of magnitude [57]. All but one of 15 triazolopyrimidine-resistant mutant cell cultures of tobacco and soybean were also cross resistant to both sulfonylureas and imidazolinones [79]. One culture was only cross resistant to sulfonylureas, but not to imidazolinones. Cross resistance of triazolopyrimidines with sulfonylurea-resistant tissue cultures of *Stellaria media* have been obtained [82] (Fig. 3). These data indicate that the molecular sites of interaction of these three chemical classes with ALS overlap, but not completely.

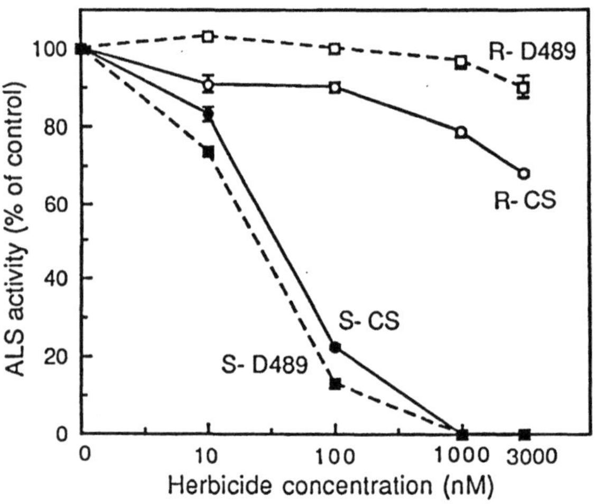

Fig. 3. In vitro effects of chlorosulfuron (CS) and the triazolopyrimidine D489 on ALS activity from resistant (R) and susceptible (S) *S. media*. From (82).

The sulfonylurea and imidazolinone binding site is separate from the feedback regulation binding sites of valine and leucine. There are no changes in the effects of valine or leucine on ALS activity by mutations that confer sulfonylurea or imidazoline resistance [49] or vice versa [52].

Weeds (*Kochia scoparia* and *Lactura serriola*) resistant to ALS-inhibiting herbicides were first reported in 1987, after only 3 to 5 years of sulfonylurea herbicide use [1, 83]. Only 88 to 123 g of sulfonylureas per hectare were used over this time period. The resistance was demonstrated to be due to altered ALS, and cross resistance to imidazolinones was found in sulfonylurea-resistant biotypes of two weed species [83]. The next year, resistant *Kochia* had spread widely and *Stellaria media*, *Ixophorus unisetus*, and *Salsolo iberica* resistance to these herbicides was first reported [1]. Earlier predictions of rapid development of weed resistance to ALS inhibitors, based on the ease of producing resistant mutants in the laboratory, appear to be correct. Biotypes of *Lolium rigidum* that rapidly metabolize aryloxyphenoxy propionate herbicides such as diclofop are resistant to both aryloxyphenoxy propionates and sulfonylureas [84].

The apparent ease with which mutant plants that are resistant to these herbicides are created and the rapidly-growing number of reports of weeds resistant to sulfonylureas and imidazolinones suggest that the ALS gene(s) is highly mutable. The complex patterns of cross resistance demonstrate that several mutations conferring resistance are possible. To date, there are few data on the biological fitness of the different ALS mutants, but all indications are that many of them are as fit as the wild types.

2.3 Glufosinate – Glutamine Synthesis

Glutamate synthetase (GS) is the last enzyme in the pathway that assimilates inorganic nitrogen into organic compounds. It is a pivotal enzyme in nitrogen metabolism in that, in addition to assimilating ammonia produced by nitrite reductase, it recycles ammonia produced by other processes, including photorespiration and deamination reactions (Fig. 4). GS is nuclear-encoded and exists in different isoforms in the cytoplasm and plastid [85, 86]. The ratio of the two isoforms varies between cell types and species, but in green tissues the chloroplast isoform is the predominant or only form found. The plastidic isoform is composed of an octamer of four or six different types of subunits. Each of the eight subunits has a reaction center [87]. Several compounds which share some structural similarities inhibit GS, however only two, glufosinate (also called phosphinothricin) and bialaphos, have been commercialized as herbicides. Bialaphos is a tripeptide (phosphinothricyl-L-alanyl-L-alanine) produced by *Streptomyces hygroscopis* which is metabolized to glufosinate and alanine by plants [88].

Glufosinate inhibits GS activity by irreversibly binding to a tetrahedral intermediate. Chloroplastic GS from spinach is inhibited by glufosinate with a K_i of 6.1 mM [89]. Inhibition of GS by glufosinate is irreversible in the presence of ATP [7]. The first step in the normal mechanism of GS activity is formation

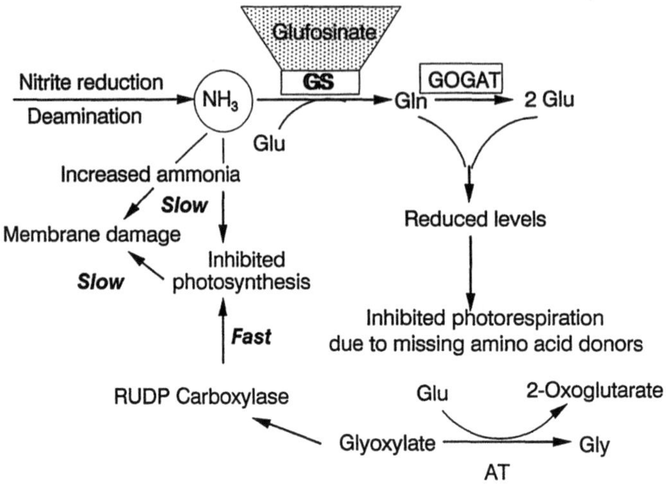

Fig. 4. Site of action of glufosinate and its effects on the physiology of the plant

of the tetrahedral adduct, λ-glutamylphosphate, from ATP and glutamate. Inactivation by glufosinate is attributed to its similar transition state analog properties when phosphorylated to the tetrahedral adduct of the substrates. Under physiological conditions, Höpfner et al. [87] labeled GS in mustard with ^{32}P-λ-glutamylphosphate adducts. Irreversible binding under these conditions led to the determination that GS contains eight reaction centers per molecule.

Inhibition of GS activity leads to a rapid accumulation of high levels of ammonia under conditions in which NO_2^- is being photosynthetically reduced and/or conditions which support photorespiration [90–93]. Similarly, barley mutants lacking GS are detected by their inability to grow under conditions that favor photorespiration [94]. Wild et al. [90] found that in mustard most of the ammonia formed as a result of glufosinate treatment was due to photorespiration rather than to nitrite reduction. Accumulation of ammonia caused by glufosinate is accompanied by cessation of photosynthesis, disruption of chloroplast structure, and vesiculation of the stroma. Although some have attributed the inhibition of photosynthesis in cells treated with GS inhibitors to effects of ammonia on photosynthesis, and on photophosphorylation in particular [e.g. 95, 96], Sauer et al. [91] found that glutamine depletion caused by glufosinate was the primary cause of inhibition of photosynthesis. Simultaneous treatment with glutamine and glufosinate resulted in little effect of glufosinate on photosynthesis, despite high ammonia levels. Exogenously supplied ammonia (up to 50 mmol per gram of fresh weight) had relatively little effect on photosynthesis. The cessation of photosynthesis caused by GS inhibitors has also been attributed to accumulation of glyoxylate (an RuBP carboxylase inhibitor) [97] and depletion of Calvin cycle intermediates [91] (Fig. 4). Suppression of the Calvin cycle results in the channeling of some energy into photosynthetic reduction of molecular oxygen, nitrite, sulfur, and phosphorus. Photoinhibition will occur, ultimately resulting in membrane lipid peroxidation via triplet chlorophyll generation and the toxic oxygen species produced by triplet chlorophyll.

Glufosinate and bialophos are non-selective herbicides. Although there is considerable variation between species in sensitivity to glufosinate, the variation is not due to differences in sensitivity of their GS [98]. Plants are tolerant or resistant to glufosinate by one of two mechanisms. In vitro selection of alfalfa cell lines with increasing concentrations of glufosinate led to tolerant (20-fold) cells whose tolerance was based on amplification of a GS-encoding gene [99]. The copy number correlated with the level of tolerance. Unfortunately, during selection, the ability to regenerate plants was lost, and cell fusion with highly regenerative lines did not restore the ability to regenerate intact plants [100].

A highly successful strategy for generating glufosinate-resistant crops was based on the protective mechanism utilized by bialophos-producing microorganisms. The glufosinate resistance *bar* gene of *Streptomyces hygroscopicus* [101] and *pat* gene of *Streptomyces viridochromogenes* [102] were found to encode for phosphinothricin (glufosinate) acetyl transferase (PAT), an enzyme that inactivates glufosinate. Considerable homology was found between the two genes. PAT has high substrate specificity for glufosinate and acetylated glufosinate is apparently herbicidally inactive and stable in vivo. Soil microorganisms can also inactivate glufosinate by transamination and oxidative deamination [103]. Transgenic, glufosinate-resistant tobacco, tomato, and potato plants have been produced by *Agrobacterium*-mediated transformation with the *bar* gene coupled with a cauliflower mosaic virus promoter [104] and similar resistant tobacco plants have been generated with the *pat* gene [102]. Transformed plants with the *bar* gene are completely resistant to standard use rates of glufosinate in the greenhouse [101] and field [105]. Glufosinate-treated resistant plants produce normal amounts of viable seed [101], and they are resistant to both glufosinate and bialophos. Three copies of the resistant gene were found to reside at a single dominant locus in transgenic, resistant plants. Rice callus has recently been transformed with the *bar* gene [106]. The resulting transformant is resistant to glufosinate.

Selection for tolerance of intact oat plants to the pathogen that produces tabtoxin, a GS-inhibiting structural analog of glufosinate, resulted in tolerant lines of oat with leaf-derived GS (both cytoplasmic and plastidic) with much less sensitivity to tabtoxin than leaf GS from nonselected oats [107]. This was the only mechanism found to account for insensitivity to the pathogen. The insensitive GS isoforms were sensitive to methionine sulfoximine, another glufosinate analog. Unfortunately, their sensitivity to glufosinate was not reported. Further work is needed to determine whether genetic manipulation of GS quantity and/or quality can confer resistance to GS-inhibiting herbicide in crops without a fitness or yield penalty.

Neither glufosinate nor bialophos have been commercially available long enough for weed resistance to have developed. The relatively short half-life of these compounds in the environment and their low soil activity should delay the occurrence of resistance.

3 Photosynthetic Inhibitors

The mechanism of action of almost half of all herbicides involves inhibition of some step in photosynthesis. Many of these compounds have been used continuously for several several decades and have a relatively long half-life in the soil. Thus, it is not surprising that most of the cases of herbicide resistance also involve resistance to herbicides that disrupt photosynthesis. Two distinct sites of action are targeted (Fig. 5). Triazines, substituted ureas, and bentazon are among the herbicides that inhibit electron flow through photosystem (PS) II. Bipyridinium herbicides, such as paraquat, accept electrons from PS I and become radical di-cations which generate superoxide radicals, causing toxic oxygen stress to the plant. Weeds have acquired resistance to both herbicide types, through two distinctly different resistance mechanisms.

3.1 Triazine Resistance

Although there was a case of 2,4-D resistance reported in 1964, the greatest impact in the area of herbicide resistance has been the discovery of triazine-resistant weeds [108]. Ryan [109] discovered a triazine-resistant (TR) biotype of *Senecio vulgaris* in the state of Washington, U.S.A. that was found in an area where only the triazine herbicides, atrazine and simazine, had been utilized as the sole herbicide treatments for several years. Weed control could not be achieved at any herbicide rate examined.

All crop plants tolerant to atrazine, such as maize, are initially affected by the herbicide but recover from the herbicide treatment by metabolism of the herbicide to inactive compounds. The triazine-resistant *Senecio* did not metabolize atrazine, however, and studies of isolated chloroplasts of the TR *Senecio* showed that electron transport through PS II was unaffected by even high herbicide levels [110]. A quick way to monitor resistance was in vivo chlorophyll fluorescence. When treated with a triazine herbicide, triazine--susceptible (TS) plants display a

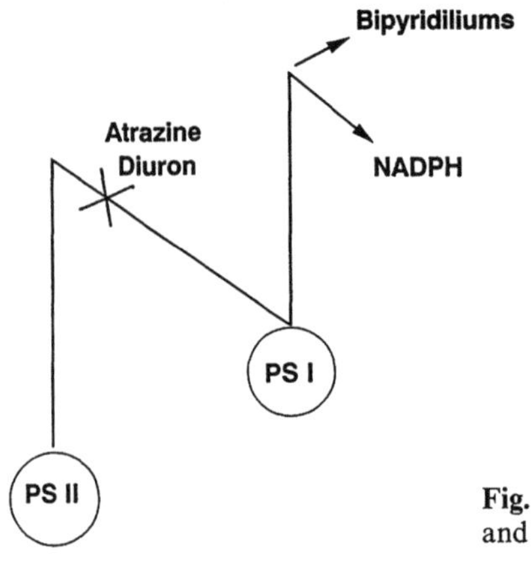

Fig. 5. Diagram of photosynthetic electron transport and the steps blocked by various herbicides

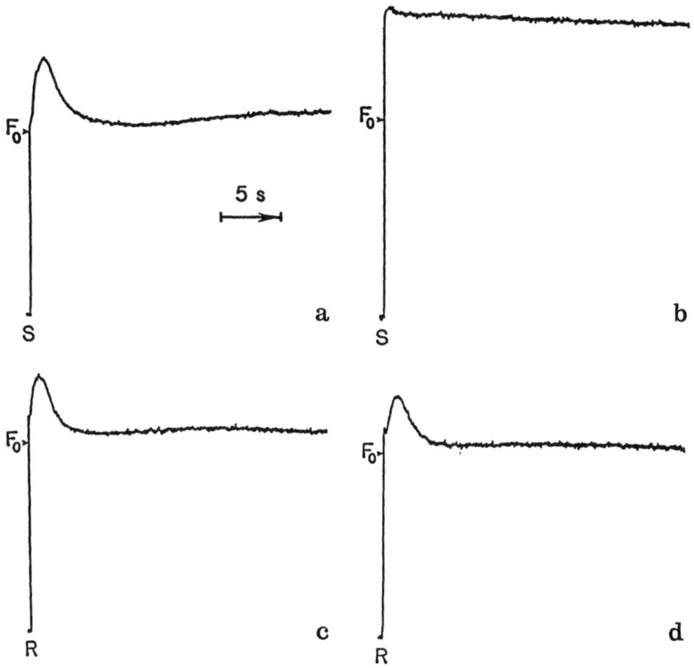

Fig. 6a–d. Fluorescence transients of untreated (**a**, **c**) and 10 μM atrazine-treated (**b**, **d**) leaf discs of TS (**a**, **b**) and TR (**c**, **d**) biotypes of *Chenopodium album*. Note that the atrazine treatment causes a rise in variable fluorescence in the S biotype but not the R

large increase in variable fluorescence, whereas fluorescence of TR plants is unaffected (111, Fig. 6).

An important step forward in this work was the identification of the protein to which atrazine binds. An analog of atrazine, azido-atrazine, forms a covalent linkage with the protein to which binding occurs [112]. When radiolabelled azido-atrazine is bound in vitro to isolated thylakoid proteins from TR and TS chloroplasts, a single band at 32 kDa is detected by fluorography, following thylakoid protein separation by polyacrylamide gel electrophoresis of the TS biotype. No comparable band is detected from the thylakoids of the TR biotype. Similarly, binding studies of C^{14}-atrazine to thylakoids reveal no specific binding of the herbicide thylakoids of the TR biotype, although there is considerable binding to thylakoids of the TS biotype. The 32 kDa protein (and triazine resistance), is maternally inherited, encoded by chloroplast DNA [113].

With the advent of DNA sequencing techniques, the changes between the *psbA* genes (the gene which encodes the 32 kDa protein) of TR and TS biotypes could be determined. Hirschberg and McIntosh [114] were the first to document a change from Ser → Gly at amino acid 264 in triazine-resistant *Amaranthus* and *Solanum* biotypes. All naturally-occurring resistant mutants which have been sequenced involve this same mutation. Other workers [115, 116] have taken advantage of the change in the nucleotide sequence and used the selective restriction enzyme *MaeI*, which recognizes a restriction site in the *psbA* sequence from TS but not TR weeds because of this nucleotide sequence change. Thus, restriction mapping of chloroplast DNA by this restriction enzyme can quickly distinguish mutations at this base.

Resistant cells selected in tissue cultures of tobacco by Japanese [116] and potato by American workers [117] indicate that, although there is a mutation in amino acid 264 in both cultures, it is not Ser → Gly but rather Ser → Thr. TR plants as well as mutants of *Chlamydomonas* contain mutations at amino acids other than 264 and differ in the level of resistance to atrazine and cross-resistance to other compounds that inhibit PS II [118].

Nearly all of the TR mutants that have been described have reduced fitness compared to the TS biotypes [74, 119, 120]. The most extensive studies in this area have been conducted by Holt and colleagues [119, 121] on TR and TS cultivars of common groundsel. In competition studies between the original resistant strain from Washington State and a cultivar from California, there was a large difference in the competitive ability of the two biotypes, with the TR less fit in every category measured. These two biotypes of common groundsel originate from completely different environments and are morphologically dissimilar. Subsequent studies by this group utilized more isogenic reciprocal F_2 and repeated backcross isolines to verify that the differences in fitness were indeed due to the triazine resistance mutation and not unrelated biotypic differences in the parental lines [121]. Similar observations have been made for TR and TS isogenic lines of canola [120, 122]. There are some reports that indicate some TR biotypes may be less affected or have some selective advantage in certain situations over the TS biotypes [123].

The basis for the loss of fitness rests chiefly in the altered electron flow through PS II. The Q_A/Q_B transition is always slower in the TR biotypes than in the TS of those weed species that have been monitored, so that electron transport through PS II is lowered [124]. It is likely that reduced electron transport through PS II sets into motion other changes associated with triazine resistance. Most of the TR biotypes investigated have larger grana stacks, relatively less stroma lamellae, much less starch, and a change in both the chlorophyll *a*/*b* ratio and the galactolipid complement ([125, 126]; Fig. 7). These shade-type chloroplast modifications indicate that the TR plants may be compensating for a lowered PS II activity by creating more PS II centers or increasing the light-harvesting ability of the centers that are present. Interestingly, a TR potato mutant with no loss of fitness also exhibits none of the changes associated with the less fit TR biotypes [117].

Although virtually all of the TR biotypes are site-of-action mutants involving amino acid 264 in the 32 kDa protein, there are at least two exceptions. A TR velvetleaf [127] recently found in Maryland is about 100-fold resistant to atrazine. Unlike the other TR biotypes investigated, however, thylakoids of this TR mutant are as sensitive as the TS biotype to inhibition of PS II activity by atrazine. These data clearly indicate that the TR velvetleaf biotype is not a site-of-action mutant. Metabolism studies indicate that the rate of metabolic degradation of atrazine, involving the enzyme glutathione S-transferase, was enhanced in the mutant

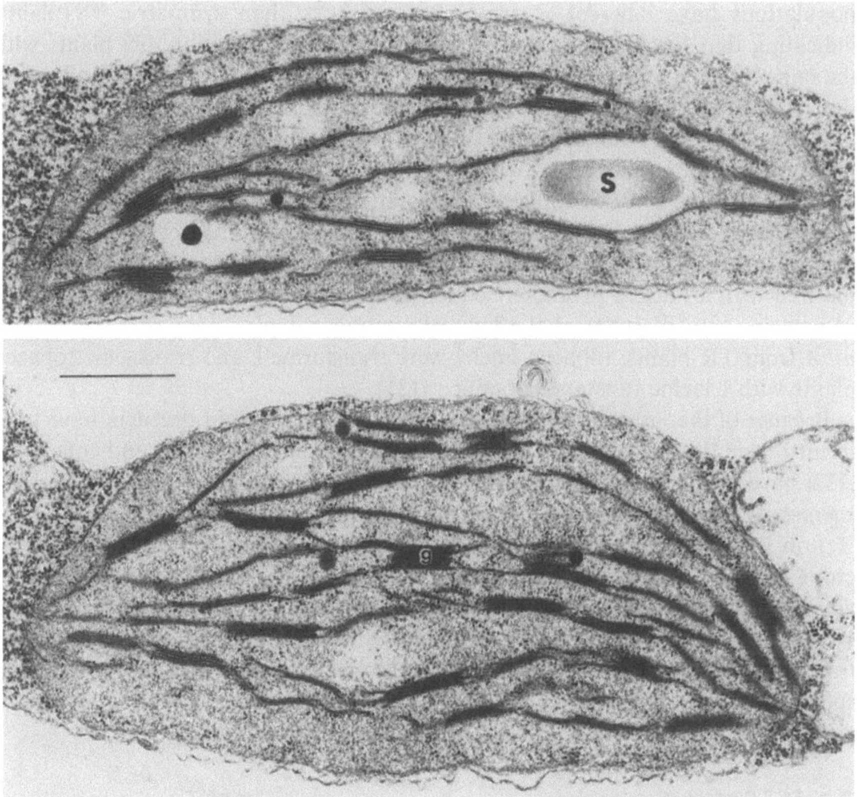

Fig. 7. Electron micrographs of TS (upper) and TR (lower) biotypes of *Brassica*. Note the larger grana (*g*) stacks and absence of starch grains (*S*) in the TR biotype. Bar = 1.0 μm

compared to the TS biotype. Although some TR biotypes that are site-of-action mutants lack enhanced rates of atrazine degradation, Gressel et al. [128] detect metabolism in a site-of-action mutant of *Brachystylon* and suggest that both enhanced metabolism and an alteration in the 32 kDa protein are present. These authors speculate that TR mutants may have been selected from populations that were initially tolerant to low (sub-lethal) levels of atrazine by enhanced metabolism.

The evolution of triazine resistance has been investigated by several groups. Arntzen and Duesing [129] describe a plastome mutator (a nuclear gene that induces alterations in chloroplast DNA) in a TR biotype of *Solanum nigrum*. These authors speculate that biotypes with plastome mutators would be more likely to evolve TR biotypes due to the enhanced mutability of the chloroplast DNA. Spontaneous chloroplast DNA mutations occur so rarely that the probability of a specific amino acid change in the 32 kDa protein would be infinitesimal without a plastome mutator. Gasquez and colleagues [130] investigated the isozyme profile populations of TR biotypes of *Chenopodium* and *Poa* and found that TR populations have different isozyme characteristics than sympatric TS plants, indicating that the TR plants arose from a founder population. TS plants with the same isozyme profile also tended to segregate out a few TR progeny, indicating that either the plant had a few copies of chloroplast DNA carrying the amino acid 264 change, but not enough to cause a TR phenotype, or these plants contain plastome mutator genes.

The *psbA* gene is a chloroplast gene, so attempts to transform plants with the altered form of the gene present in TR plants have been unsuccessful because of the low success in chloroplast transformation. However, by constructing a chimeric plasmid with a nuclear promoter sequence, a transit peptide (necessary for entry of proteins into plastids) of the light-harvesting complex gene and the mutant *psbA* from TR plants, tobacco nuclei were transformed, and transgenic tobacco plants with triazine resistance obtained [131].

Because of the large number of TR species available, plant breeders have tried to introduce the TR factor by classical methods. Souza-Machado and colleagues [132] have been successful in transferring the TR cytoplasm of the wild *Brassica campestris* into a number of other *Brassica* spp. One of these, the rapeseed variety 'Triton' is marketed in Canada. Widespread use and acceptance of triazine-resistant crops remain limited because of yield reductions of 15 to 25% due to the previously described loss in plant fitness. A similar series of crosses have transferred cytoplasm of a TR *Setaria viridis* into *Setaria italica* [133]. These kinds of genetic transfer are possible only when there are sexually compatible weed and crop species.

3.2 Paraquat Resistance

Similar to the first cases of triazine resistance, the first cases of paraquat resistance occurred in areas where paraquat was the only herbicide applied [134, 135]. Furthermore it was applied repeatedly during a single growing season. The genera with the most reports of resistance [136] are the two closely related genera *Conyza* and *Erigeron*, where resistant biotypes have been noted world-wide, and *Hordeum*, where two Australian species have been noted as paraquat-resistant (PR). The level of resistance to paraquat is greater than 100-fold in all of these cases, although there is some low level ($>$10-fold) resistance to paraquat in two grass species in the United Kingdom [137, 138].

Two theories have been put forth to explain the mechanism(s) of resistance in these PR biotypes: (a), detoxification of toxic oxygen species by enhanced activities of enzymes in the Halliwell-Asada pathway [139]; and (b), sequestration of paraquat away from its chloroplastic site of action [140]. Other potential mechanisms for resistance have been discussed by Fuerst and Vaughn [136]; these other mechanisms are not considered here because none of the resistant biotypes have these resistance mechanisms.

The first published report on the PR *Conyza* was by Youngman and Dodge [134]. These workers reported that a novel isozyme of the chloroplastic Cu–Zn superoxide dismutase (SOD) was found in the paraquat-sensitive (PS) biotype. Moreover, a 300% increase in total SOD activity was found in the PR biotype. These workers concluded that the increased SOD activity and the novel isozyme was the cause of increased resistance. Vaughn and Fuerst [141] used an F_2 population from a cross of PR and PS *Conyza* biotypes that segregated 3:1 for the resistance trait. Both parental biotypes and the segregating progeny had identical isozyme profiles for all three SOD isozymes, regardless of their paraquat resistance [141]. Shaaltiel and colleagues [139, 142] found the same isozyme profile in their biotypes in total homogenates although they observed a 120% increase in SOD activity if chloroplast extracts of the two biotypes were compared. This relatively small increase (compared to the relatively large increase in paraquat resistance) could be due to the differences in the ability to isolate intact chloroplasts (since SOD is a soluble stromal enzyme) from the two biotypes or to the differences in enzyme deterioration because of the higher phenolic content of the PS biotype (Vaughn, unpublished). No increase and/or isozyme changes have been detected in a number of other PR-resistant biotypes [143–145].

There are two serious objections to the role of SOD or other enzymes in the mechanism of paraquat resistance in *Conyza*. If SOD is involved in the mechanism of paraquat resistance, then one would expect similar levels of cross resistance to other compounds that cause the generation of toxic oxygen species. However, Vaughn et al. [146] found that, although there was a 100-fold resistance to paraquat, there was only a 10-fold resistance to diquat, and no resistance to morfamquat or to several other toxic oxygen generators. Another serious objection lies in the mechanism of action of paraquat itself. Paraquat continually generates toxic oxygen radicals. There would also be no energy to generate reductant, as paraquat would intercept electrons from the reducing side of PS I, preventing NADPH production and re-reduction of glutathione or ascorbate. Thus, even if a protective system were present, it would ameliorate the effects of paraquat for only a short time. The PR *Conyza* and other biotypes were selected under field conditions, based upon their ability to withstand paraquat over long periods and through repeated applications [134, 135].

Considerable evidence exists to support the view that paraquat is sequestered at some site away from the chloroplast. Fuerst et al. [140] and Vaughn et al. [146] found that the ability of paraquat to suppress chlorophyll fluorescence required a 100-fold higher concentration in the PR than the PS *Conyza* biotype. Other compounds that had a similar mode of action to paraquat either did not show any difference (e.g. morfamquat) or a much smaller difference (e.g. diquat). Autoradiographic studies of Fuerst et al. [140], Tanaka et al. [147], and Bishop et al. [148] all showed that the movement of ^{14}C-paraquat was much more restricted in the PR than in the PS biotypes. Some of these studies were performed in darkness so that the effects of paraquat damage to membranes (and hence the more unrestricted movement of the herbicide) would be minimized. However, studies of Tanaka et al. [147] utilized both illuminated and non-illuminated conditions for labelling and both revealed a restricted pattern of paraquat distribution in the PR but not the PS biotypes. The mechanisms by which paraquat is sequestered in these biotypes is not known. Vaughn et al. [146] speculated that

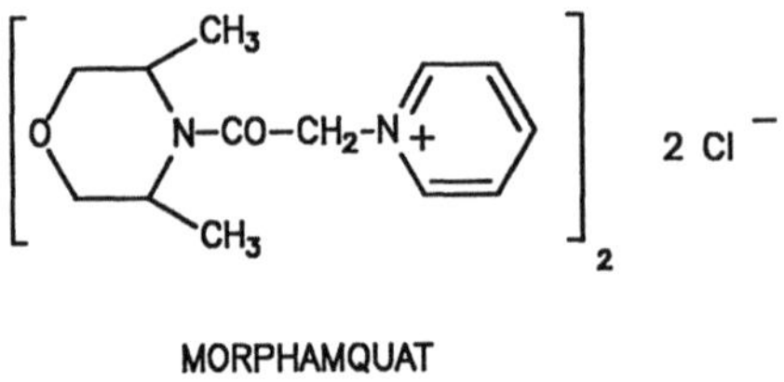

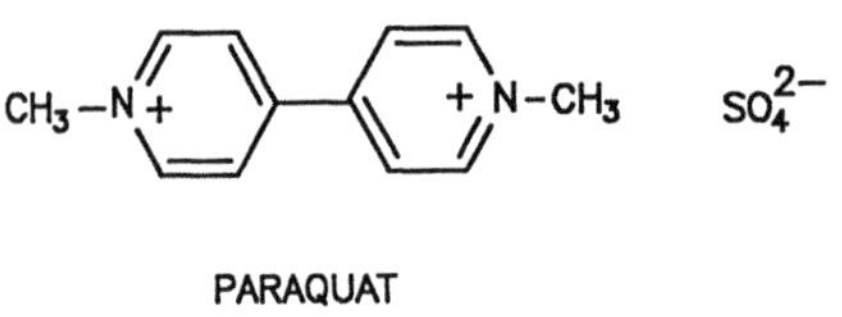

Fig. 8. Structures of paraquat and morfamquat. The PR biotype of *Conyza* his highly resistant to paraquat, but exhibits no cross-resistance to morfamquat despite the fact that both compounds have an identical mechanism of action

the mechanism of paraquat resistance involved binding interactions with the charged ends of the paraquat molecule because morfamquat is structurally similar to paraquat but has bulky side groups that surround the charged nitrogens(Fig. 8).

Whatever resistance mechanism one ascribes to, there is a consensus of the genetic data on the PR biotypes. In both *Erigeron* and *Conyza* biotypes, the paraquat resistance trait is controlled by a monogenic dominant gene [149] whereas in *Hordeum* [150] and the fern *Ceratopteris* [151], the trait is controlled by a monogenic recessive gene. As work progresses in this area, it will be interesting to determine if the molecular sites of these various PR biotypes are similar or whether different proteins are responsible for the resistance phenotype. Unlike the TR mutants, the PR mutants do not appear to be at a selective disadvantage, although careful studies, such as performed by Holt [119] with TR mutants, have not been performed.

4 Mitotic Inhibitors – Dinitroaniline Resistance

The appearance of dinitroaniline-resistant (DR) biotypes of several weed species is not unexpected. Dinitroaniline herbicides, such as trifluralin and pendimethalin, have been used extensively (and often exclusively) for weed control in cotton in the Southeastern United States for a number of years and the degree of control of susceptible species is often greater than 90% [152]. Both of these conditions (repeated use, high mortality rate) greatly enhance the selection pressure for DR biotypes by eliminating dinitroaniline-susceptible (DS) biotypes from the population.

Dinitroaniline herbicides affect the ability of tubulin subunits to polymerize into microtubules [153, 154]. As a consequence, cellular processes that require microtubules such as chromosome movement during mitosis, cell shape, and cell plate formation are disrupted so that neither cell division nor elongation may occur. In rapidly growing and elongating tissues such as root tips, typical symptoms include swollen, clubby roots [154].

Studies of the DR *Eleusine* indicated that this biotype exhibited no root tip swelling at 1 000–10 000 times the herbicide level that caused root tip swelling in the DS biotype [155]. More extensive studies, utilizing a number of herbicides that disrupt mitosis, reveal that the DR biotype is cross-resistant to all of the dinitroaniline herbicides and the unrelated herbicide amiprophosmethyl [156]. Even at saturated aqueous concentrations of the herbicides, microtubules are present in the DR biotype.

Biochemical investigations indicate that the DR *Eleusine* is altered at the site of action of the herbicide, the tubulin protein. Tubulin isolated from both biotypes polymerizes into microtubules in the absence of dinitroaniline herbicide whereas only the tubulin of the R biotype makes microtubules in the presence of herbicide [157]. Western blots probed with polyclonal antisera that recognize both α and β tubulin (Fig. 9) as well as monoclonal antisera to β tubulin indicate that the R biotype has a second novel form of tubulin not found in the S biotype. In addition, the DR biotype appears to make more tubulin protein per gram fresh weight or per mg protein than the DS. Southern hybridization with *Arabidopsis* tubulin probes indicates that the DR biotype also contains four copies of the β-tubulin gene not found in the DS (Marks and Weeks, unpublished).

The mechanism of resistance appears to be more complex than a simple difference in tubulin that reduces the affinity of the herbicide. Animal cell lines resistant to microtubule-disrupting compounds have hyperstabilized microtubules (that is, microtubules that remain in the polymerized form, microtubules, rather than as free tubulin heterodimers) and are hypersensitive to compounds that stabilize

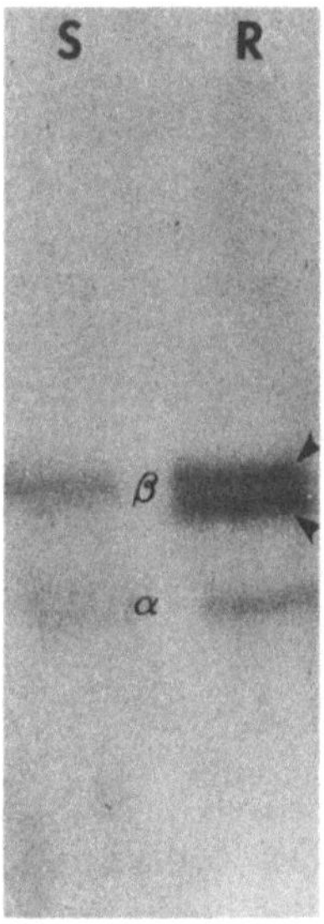

Fig. 9. Western blot of root tip extracts from 50 seedlings of DR (*R*) and DS (*S*) biotypes of *Eleusine* probed with a polyclonal antitubulin. The *R* biotype has more tubulin protein (as evidenced by heavier staining) and a doublet of β tubulin (*arrow*) rather than a single form. These blots were from gel separations that maximized separation of α and β tubulin subunits and are not reflective of molecular weights of these proteins

microtubules such as taxol. Similarly, the DR *Eleusine* is hypersensitive to taxol, with the DR biotype about 50 to 100-fold *more* sensitive than the DS biotype, by a number of criteria [157, 158]. Thus, it appears that the alteration in β-tubulin confers hyperstability to the microtubules, allowing them to remain polymerized (or perhaps even increase in length) in the presence of herbicide.

Much less is known about the other DR biotypes. A novel biotype of *Eleusine* with less resistance than the DR biotype was also discovered in certain counties in South Carolina, USA [152]. It has been termed the intermediate resistant (IR) biotype. The IR biotype differs from the DR not only in the level of resistance but also in the patterns of cross-resistance to other dinitroaniline herbicides. The IR biotype is highly resistant to isopropalin, moderately resistant (50 to 100×) to most dinitroanilines and amiprophosmethyl, and only sparingly (5 to 10×) resistant to oryzalin [159], based on the concentrations of the herbicides required to disrupt mitosis. The IR biotype is not markedly more sensitive to taxol than the DS [158]. The mechanism of resistance in this biotype appears to be different from the DR, but the exact molecular mechanism is not known.

Even more recently, a DR biotype of *Setaria viridis* was found in Canada [150]. Preliminary investigations in our laboratory indicate that this biotype has a much lower level of resistance (2 to 10×) than the DR biotype of *Eleusine*. This biotype is cross-resistant to all the other dintroaniline herbicides as well as the herbicides dithiopyr, amiprophos-methyl, and very highly resistant (>100×) to DCPA [161]. This is the first report of a DCPA-resistant biotype. The DR *Setaria* is also hypersensitive to the herbicide chlorpropham, as is the DR *Eleusine* [157, 161]. Again, nothing is known about the mechanism of resistance.

Fitness studies of non-isogenic lines of the DR and DS *Eleusine* indicate that there is little loss of vegetative fitness associated with dinitroaniline resistance [162]. Morphologically the two biotypes appear no different. Only seed dry weight (a measure of reproductive fitness) was reduced in the DR biotype.

Compared to the simple genetics of the TR and PR biotypes, the inheritance of dinitroaniline resistance is quite complex. F_1 hybrids between the DR and DS biotypes are resistant, although the DR F_1 hybrids exhibit some root tip swelling at the highest (10 μM) herbicide concentrations, unlike the DR parent. The F_2 exhibits a wide range in the level of root tip swelling, with some plants comparable to the DR parent while others are fully as susceptible as the DS parent (Vaughn, unpublished). The presence of four new copies of the β-tubulin gene in the DR plant (Marks and Weeks, unpublished) would indicate a complex inheritance pattern if the copies were not tandemly linked.

5 Resistance to Other Herbicides

5.1 MCPA and 2,4-D Resistance

The herbicides MCPA and 2,4-D are used extensively and effectively in pasture settings to control broadleaved weeds. Recently, farmers in New Zealand have found these herbicides to be less effective than before in controlling both *Carduus nutans* and *Ranunculus acris*. Resistance of new biotypes has been verified in greenhouse studies of these putative resistant populations [163]. The scenario for the appearance of resistance is much the same as for the other resistant biotypes: 2,4-D or MCPA were used annually, without rotation with other herbicides, for the control of these weeds for up to 25 years. In some cases, no alternative herbicide (e.g. clopyralid) is possible because the pastures contain both legumes and pasture grasses.

The MCPA-resistant *Carduus* is cross-resistant to MCPB, 2,4-D, and the sulfonylurea DPX-L5300, but not to mecoprop, clopyralid, dicamba, picloram, paraquat, or glyphosate [164]. These results are indicative of a resistance based upon metabolism, but, as for the multiple resistances in *Lolium* (see below), there may be several operative mechanisms for this pattern of cross-resistance. Fitness studies indicate that the MCPA-resistant *Carduus* is not significantly lower in fitness than susceptible populations. This would also indicate that these biotypes are not altered at the site of action, based upon other herbicide-resistance studies.

5.2 Resistance to Herbicides that Affect Cell Wall Formation

Both dichlobenil and isoxaben (a herbicide under development by Dow-Elanco) disrupt cell plate and secondary cell wall formation in higher plants [165, 166]. Although the molecular site of action for these herbicides is not known, and resistant biotypes have not been found under field conditions, resistant mutants from both cell cultures of tomato and whole plants of *Arabidopsis* have been obtained [167–169].

Delmer [167] isolated tissue culture mutants that were highly resistant to dichlobenil. The mutants have abnormal cell wall complements, with a 250% increase in uronic acid and a 75% reduction in cellulose and xyloglucan. Thus, the abnormal cell walls in these resistant mutants are indicative of a site of action mutation in a process involving cell wall formation.

Several mutants of *Arabidopsis* have been isolated that are resistant to high levels of isoxaben. The mutants each appear to be single-gene recessives (two different loci are involved) but the identity of the gene products or the mechanism for the resistance has not been elucidated.

6 Multiple Resistance

In the other herbicide-resistant mutants investigated to date, there has been some cross-resistance to other herbicide groups, generally to with the same mechanism of action or structure. For example, PR biotypes are cross-resistant to diquat [146]. Multiple resistant (MR) biotypes have been noted both in Australia and England that are not only resistant to the herbicide which selected for them, but also other herbicides with different mechanisms of action.

The best-studied example of multiple resistance is the MR biotype of *Lolium rigidum* (reviewed in 170, 171]. This biotype was originally selected as a diclofop-methyl resistant mutant but cross-resistance studies indicated that this biotype was resistant not only to those grass herbicides that affect acetyl Co–A carboxylase (ACC) but also to sulfonylureas, imidazolinones, and some dinitroanilines. Biotypes from different localities appear to have different cross-resistance patterns, some with broad cross-resistance and others with relatively narrow patterns of cross-resistance. This extensive cross-resistance is particularly alarming in that it leaves almost no herbicides that could effectively control these weeds. The actual mechanism(s) for multiple resistance has not been determined, although some possibilities have been eliminated [170, 171]. Neither ACC nor ALS, the sites of action of the aryloxyphenoxypropionates and the sulfonylureas, respectively, were different between the MR biotypes and their susceptible counterparts. Likewise, there was no significant difference in uptake and/or translocation of the herbicides in the two biotypes. Thus, the mechanisms of resistance found in the biotypes of weeds resistant to only one class of herbicide are not the basis for resistance in MR *Lolium*.

Herbicide metabolism was the next logical mechanism for examination. Because many insecticides are metabolized by mixed function oxidases (MFOs), inhibitors of MFO's such as aminobenzotriazole were used to suppress the MFOs of *Lolium* to test this mechanism of resistance. Combination treatments of MFO inhibitors and herbicide on MR *Lolium* and *Alopecurus* [172] resulted in some lowering of the levels of resistance to these herbicides, indicating that MFOs are at least partially responsible for the MR.

The measurement of metabolites in some cases gives evidence in favor of accelerated herbicides degradation. Metabolism of both chlorsulfuron [170, 173] and chlortoluron [173] are enhanced in both *Lolium* and *Alopecurus* MR and the products are typical of MFO-type oxidation. However, diclofop-methyl metabolism is not significantly different between the MR biotypes and their susceptible counterparts [171]. It is possible that some alternative factor, such as a change in membrane structure, could influence metabolite production in the case of one herbicide but still confers resistance to another because of a subcellular compartmentalization difference for one herbicide and enhanced metabolism for another. For example, the endoplasmic reticulum is the location of cytochrome P450 in higher plants and an increase in this organelle might allow for both metabolism and compartmentalization.

As mentioned above, the biotypes differ greatly in their patterns of cross-resistance and it is possible that the multiple resistance of the *Lolium* biotypes is not due to a single gene difference. Genetic studies on this group, which have just begun, will be valuable in determining whether single or multiple mechanisms (and genes) are involved.

7 Prospectus

Much is still unknown in the cases of herbicide resistance investigated to date. The TR biotypes are the most thoroughly investigated, but the production of biotypes with better fitness (alternate mutations at amino acid 264), is an exciting development in this area. The two proposed mechanisms of paraquat resistance (detoxification of toxic oxygen vs. compartmentalization) must be reconciled. Investigations of DR biotypes of *Setaria* are just beginning; these mutants apparently do not have the same mechanism or cross-resistance pattern as the DR *Eleusine*. The rapid appearance of biotypes resistant to ALS inhibitors, predicted by early mutation studies, will be a ripe source of many different mutations of the ALS protein. The study of MR biotypes is the most exciting recent development in the area of herbicide resistance. Elucidation of the mechanisms for resistance and designing herbicide strategies to prevent its occurrence will be a major challenge for weed scientists.

8 References

1. McFarland J, LeBaron HM (1990) Amer Chem Soc Symp Ser 421: 336
2. Mazur BJ, Falco SC (1989) Annu Rev Plant Physiol Plant Mol Biol 40: 441
3. Conn EE (ed) (1986) The Shikimate Pathway. Plenum, New York (Rec Adv Phytochem, vol 20)
4. Morris PF, Doong RL, Jensen RA (1976) Plant Physiol 89: 10
5. Duke SO, Naylor AW (1976) Plant Sci Lett 6: 361
6. Homeyer U, Schultz G (1988) Plant Physiol Biochem 26: 365
7. Kishore GM, Shah DM (1988) Annu Rev Biochem 57: 627
8. Duke SO (1988) In: Kearney PC, Kaufman DD (eds) Herbicides – chemistry, degradation, and mode of action. Marcel Dekker, New York, pp 1–70
9. Cole DJ (1985) In: Grossbard E, Atkinson D (eds) The herbicide glyphosate, Butterworths, London, pp 48–74
10. Ream JE, Steinrücken HC, Porter CA, Sikorski JA (1988) Plant Physiol 87: 232
11. Padgette SR, Huynh QK, Aykent S, Sammons RD, Sikorski JA, Kishore GM (1988) J Biol Chem 263: 1798
12. Anderson KS, Sikorski JA, Johnson KA (1988) Biochemistry 27: 1604
13. Anderson KS, Sikorski JA, Johnson KA (1988) Biochemistry 27: 7395
14. Padgette SR, Smith CE, Huynh QK, Kishore GM (1988) Arch Biochem Biophys 266: 254
15. Huynh QK (1988) J Biol Chem 262: 11631
16. Huynh QK, Bauer SC, Bild GS, Kishore GM, Ganesh M, Borgmeyer JR (1988) J Biol Chem 263: 11636
17. Stalker DM, Hiatt WR, Comai L (1985) J Biol Chem 260: 2724
18. Shah DM, Horsch RB, Klee HJ, Kishore GM, Winter JA, Tumer NE, Hironaka CM, Sanders PR, Gasser CS, Aykent S, Siegel NR, Rogers SG, Fraley RT (1986) Science 233: 478
19. Holländer-Czytko H, Amrhein N (1987) Plant Physiol 83: 229
20. Gasser CS, Winter JA, Hironaka CM, Shah DM (1988) J Biol Chem 263: 4280
21. Della-Cioppa G, Bauer C, Klein B, Shah D, Fraley R, Kishore G (1986) Proc Natl Acad Sci USA 83: 6873
22. Della-Cioppa G, Kishore GM (1988) EMBO J 7: 1299
23. Goldsbrough PB, Hatch EM, Huang B, Kosinski WG, Dyer WE, Herrman KM, Weller SC (1990) Plant Sci 71: 53
24. Nafziger ED, Widholm JM, Steinrücken HC, Killmer JL (1984) Plant Physiol 76: 571
25. Smith CM, Pratt D, Thompson GA (1986) Plant Cell Rep 5: 298
26. Smart CC, Johanning D, Muller G, Amrhein N (1985) J Biol Chem 260: 16338
27. Steinrücken HC, Schulz A, Amrhein N, Porter CA, Fraley R (1986) Arch Biochem Biophys 244: 169

28. Holländer-Czytko H, Johanning D, Meyer HE, Amrhein N (1988) Plant Molec Biol 11: 215
29. Smart CC, Amrhein N (1987) Planta 170: 1
30. Rogers SG, Brand LA, Holder SB, Sharps ES, Bracklin MJ (1983) Appl Environ Microbiol 46: 37
31. Klee HJ, Muskopf YM, Gasser CS (1987) Mol Gen Genet 210: 437
32. Jordan MC, McHughen A (1988) Plant Cell Rep 7: 281
33. Sost D, Schulz A, Amrhein N (1984) FEBS Lett 173: 238
34. Kishore GM, Brundage L, Kolk K, Padgette SR, Rochester D, Huynh QK, Della-Cioppa D (1986) Fed Proc Amer Soc Expl Biol 45: 1506
35. Della-Cioppa G, Bauer SC, Taylor ML, Rochester DE, Klein BL, Shah DM, Fraley RT, Kishore GM (1987) Bio/Technology 5: 579
36. Padgette SR, Della-Cioppa G, Shah DM, Fraley RT, Kishore GM (1989) in: Cell Culture and Somatic Cell Genetics of Plants, Vol 6. Academic, New York p 441
37. Comai L, Facciotti D, Hiatt WR, Thompson G, Rose RE, Stalker DM (1985) Nature 327: 741
38. Fillatti JJ, Kiser J, Rose R, Comai L (1987) Bio/Technol 5: 726
39. Riemenschneider DE, Haissig BE, Sellmer J, Fillatti JJ (1988) in: Ahuja MR (ed) Somatic Cell Genetics of Woody Plants, Kluwer, Boston, pp 73–80
40. Fillatti JJ, Haissig BE, McCown B, Comai L, Riemenschneider D (1988) in: Hanvoer JW, Keathley DE (eds) Genetic Manipulation of Woody Plants, Plenum, New York, pp 243–249
41. Thompson GA, Hiatt WR, Facciotti D, Stalker DM, Comai L (1987) Weed Sci 35: (Suppl 1) 19
42. Comai L, Larson-Kelly M, Kiser J, Mau CJD, Pokalsky AR, Shewmaker CK, McBride K, Jones A, Stalker DM (In Press) J Biol Chem
43. Larson-Kelly N, Comai L, Kiser J, Mau C, Pokalsky AR, McBride K, Jones A, Shewmaker C, Stalker DM (1988) SAAS Bull 1: 37
44. Benfey PN, Chua NH (1989) Science 244: 174
45. Moore JK, Braymer HD, Larson AD (1983) Appl Environ Microbiol 46: 316
46. Jacob GS, Garbow JR, Hallas LE, Kimack NM, Kishore GM, and Shafer J (1988) Appl Environ Microbiol 54: 2953
47. Beyer EM Jr, Duffy MJ, Hay JH, Schleuter DD (1988) Sulfonylureas. In: Herbicides – Chemistry, Degradation, and Mode of Action, PC Kearney, DD Kaufman, (eds), Dekker, New York, pp 117–189
48. Smith JE, Mauvais CJ, Knowlton S, Mazur BJ (1988) 379: 25
49. Singh BK, Newhouse KE, Stidham MA, Shaner DL (1989) Brit Crop Protect Conf Monograph Ser 42: 87
50. Bryan JK (1980) In: The Biochemistry of Plants, A Comprehensive Treatise. Vol 5. Amino Acid and Derivatives. BF Miflin, (ed), Academic, New York, pp 403–452
51. Durner J, Böger P (1988) Z Naturforsch 43c: 850
52. Singh BK, Stidham MA, Shaner DL (1988) J Chromatog 444: 251
53. Creason GL, Chaleff RS (1988) Theor Appl Genet 76: 177
54. Chaleff RS, Mauvais CJ (1984) Science 224: 1443
55. Chaleff RS, Bascomb NF (1987) Mol Gen Genet 210: 33
56. Mazur BJ, Chui C-F, Smith JK (1987) Plant Physiol 85: 1110
57. Schloss JV, Ciskanik LM, Van Dyk DE (1988) Nature 331: 360
58. Hawkes TR (1989) Brit Crop Protect Conf Mongraph 42: 131
59. LaRossa RA, Schloss JV (1984) J Biol Chem 259, 8753
60. Chaleff RS, Ray TB (1984) Science 223: 1148
61. Muhitch MJ, Shaner DL, Stidham MA (1987) Plant Physiol 83: 451
62. LaRossa RA, Smulski DR (1984) J Bacteriol 160: 391
63. Shaner DL, Anderson PC, Stidham MA (1984) Plant Physiol 76: 545
64. Hawkes TR, Howard JL, Pontin SE (1989) In: Herbicides and Plant Metabolism, Dodge AD (ed), Cambridge Univ, Press, Cambridge, pp 113–136
65. Jordan MC, Mchugen A (1987) J Plant Physiol 131: 333

66. Friedberg D, Seiffers J (1988) Arch Microbiol 150: 278
67. Winder T, Spaulding MH (1988) Mol Gen Genet 213: 394
68. Harnett ME, Newcomb JR, Hodson RC (1987) Plant Physiol 85: 898
69. Swanson EB, Coumans MP, Brown GL, Patel JD, Beversdorf WD (1988) Plant Cell Rep 7: 83
70. Lee KY, Townsend J, Tepperman J, Black M, Chui CF, Mazur B, Dunsmir P, Bedbrook P (1988) EMBO J 7: 1241
71. Haughn GW, Somerville CR (1986) Mol Gen Genet 204: 430
72. Haughn GW, Smith J, Mazur B, Somerville C (1988) Mol Gen Genet 211: 266
73. Olszewski NE, Martin FB, Ausubel FM (1988) Nucl Acids Res 16: 10785
74. Sebastian SA, Chaleff RS (1987) Crop Sci 27: 948
75. Sebastian SA, Fader GM, Ulrich JF, Forney DR, Chaleff RS (1990) Crop Sci 29: 1403
76. Shaner DL, Anderson PC (1985) In: Zaitlin M, Day PR, and Hollaender A, (eds), Biotechnology in Plant Science. Academic, NY, pp 287–300
77. Swanson EB, Herrgesell MJ, Arnoldo M, Sippell DW, Wong RSC (1989) Theor. Appl Genet 78: 525
78. Subramanian MV, Gerwick BC (1989) Amer Chem Soc Symp Ser 389: 277
79. Subramanian MV, Pao Brit L (1989) Pest Control Conf Mongraph Ser 42: 97
80. Saxena PK, King J (1988) Plant Physiol 86: 863
81. Gabard JM, Charest PJ, Iyer VN, Miki BL (1989) Plant Physiol 91: 574
82. Hall L, Devine M (1990) Plant Physiol 93: 962
83. Thill DC, Mallory CA, Saari LL, Cotterman JC, Primiani MM (1989) WSSA Abst 29: 132
84. Powles SB, Holtum JAM, Matthews JM, Liljegren DR (1990) Amer Chem Soc Symp Ser 421: 394
85. Botella JR, Verbelen JP, Valpuesta V (1990) Plant Physiol 88: 943
86. McNally S, Hirel B, Gadal P, Mann AF, Stewart GR (1990) Plant Physiol 72: 22
87. Höpfner M, Reifferscheid G, Wild A (1988) Z Naturforsch 43c: 194
88. Wild A, Zeigler C (1989) Z Naturforsch 44c: 97
89. Ericson MC (1985) Plant Physiol 79: 923
90. Wild A, Sauer H, Rühle W (1987) Z Naturforsch 42c: 263
91. Sauer H, Wild A, Rühle W (1987) Z Naturforsch 42c: 270
92. Rhodes D, Deal L, Haworth P, Jamieson GC, Reuter CC, Ericson MC (1986) Plant Physiol 82: 1057
93. Turner JG (1981) Physiol Plant Pathol 19: 57
94. Wallsgrove RM, Turner JC, Hall NP, Kendall AC, Bright SWJ (1987) Plant Physiol 83: 155
95. Turner JG, Taha RR, Debbage J (1986) Physiol Plant 67: 649
96. Franz TA, Peterson DM, Durbin RD (1982) Plant Physiol 69: 345
97. Ikeda M, Ogren WL, Hageman RH (1984) Plant Cell Physiol 25: 447
98. Ridley SM, McNally SR (1985) Plant Sci 39: 31
99. Donn G, Tischer E, Smith JA, Goodman HM (1984) J Molec Appl Genet 2: 621
100. Deak M, Donn G, Feher A, Dudits D (1988) Plant Cell Rep 7: 158
101. Thompson CJ, Movva NR, Tizard R, Crameri R, Davies JE, Lauwerys M, Botterman J (1987) EMBO J 6: 2519
102. Wohlleben W, Arnold W, Broer I, Hillemann D, Strauch E, Pühler A (1988) Gene 70: 25
103. Bartsch K, Tebbe CC (1989) Appl Environ Microbiol 55: 711
104. de Block M, Botterman J, Vandewiele M, Dockx J, Thoen C, Gossele V, Movva NR, Thompson C, van Montagu M, Leemans J (1987) EMBO J 6: 2513
105. De Greef W, Delon R, De Block M, Leemans J, Botterman J (1989) Bio/Technol 7: 61
106. Dekeyser R, Claes B, Marichal M, Van Montagu M, Caplan A (1989) Plant Physiol 90: 217
107. Knight TJ, Bush DR, Langston-Unkefer PJ (1988) Plant Physiol 88: 333
108. LeBaron HM, Gressel J (1982) Herbicide resistance in plants, John Wiley, New York
109. Ryan GF (1970) Weed Sci 18: 614
110. Radosevich SR, DeVillers OT (1976) Weed Sci 24: 229

111. Ducruet JM, Gasquez J (1978) Chemosphere 8: 691
112. Pfister K, Steinback KE, Gardner G, Arntzen CT (1981) Proc Natl Acad Sci USA 781
113. Souza Machado V, Bandeen JD, Stephenson GR, Lavigne D (1978) Can J Plant Sci 58: 977
114. Hirschberg J, McIntosh L (1983) Science 22: 1346
115. Darmency H, Gasquez J (1990) Amer Chem Soc Symp Ser 421: 353
116. Shigematsu Y, Sato F, Yamado Y (1989) Plant Physiol 89: 986
117. Smeda RJ (1990) PhD Dissertation, Purdue, Univ, W Lafayette, IN
118. Erickson JM, Rahive M, Rochaix JD, Mets L (1988) Science 228: 204
119. Holt JS (1990) Amer Chem Soc Symp Ser 421: 419
120. Gressel J, Ben-Sinzi G (1985) Plant Sci 38: 29
121. Stowe AE, Holt JS (1988) Plant Physiol 87: 183
122. Vaughn KC (1986) Plant Physiol 82: 859
123. Schonfield M, Yaacoby T, Michael O, Rubin B (1987) Plant Physiol 83: 329
124. Holt JS, Radosevich SR, Stemler AJ (1983) Biochim Biophys Acta 722: 245
125. Burke JJ, Wilson RF, Swafford JR (1982) Plant Physiol 70: 24
126. Vaughn KC, Duke SO (1984) Physiologia Plant 62: 182
127. Gronwald JW, Andersen RN, Yee C (1991) Proc Weed Sci Soc Amer 31: 49
128. Gressel J, Shimabukuro RH, Ompeu ME (1983) Pestic Biochem Physiol 19: 361
129. Arntzen CJ, Duesing JH (1982) In: Ahmed F, Schultz J, Voellony RW (eds), Advances in gene technology, molecular genetics of plants and animals, Academic, New York, p 273
130. Al Mouemar A, Gasquez J (1983) Weed Res 23: 141
131. Cheung AY, Bogorad L, Van Montagu M, Schell J (1988) Proc Natl Acad Sci USA 85: 391
132. Ali A, Fuerst EP, Arntzen CJ, Souza Machado F (1986) Plant Physiol 80: 511
133. Darmency HM, Pervis H (1985) Weed Res 25: 175
134. Youngman RJ, Dodge AD (1981) In: Akoyunoglou G (ed) Photosynthesis and the environment, Balaban Int Sci Serv, Philadelphia, p 537
135. Itoh K, Miyahara M (1985) Proc Tenth Asian Pac Weed Conf 13
136. Fuerst EP, Vaughn KC (1990) Weed Tech 43: 150
137. Harvey DB, Harvey BMR (1978) Plant Cell Envir 1: 211
138. Faulkner JS (1974) Euphytica 23: 281
139. Shaaltiel Y, Glazer A, Bocion P, Gressel J (1988) Pestic Biochem Physiol 31: 13
140. Fuerst EP, Nakatani HY, Dodge AD, Penner D, Arntzen CJ (1985) Plant Physiol 77: 984
141. Vaughn KC, Fuerst EP (1985) Pestic Biochem Physiol 24: 86
142. Shaaltiel Y, Gressel J (1986) Pestic Biochem Physiol 26: 22
143. Carroll EW, Schwarz OJ, Hickok LG (1988) Plant Physiol 87: 651
144. Powles SB, Cornic G (1987) Aust J Plant Physiol 14: 81
145. Hirata T, Matsanaka S (1985) Weed Res (Japan) 30: 127
146. Vaughn KC, Vaughn MA, Cammileri P (1989) Weed Sci 37: 5
147. Tanaka Y, Chisaka H, Saka H (1986) Physiol Plant 66: 605
148. Bishop T, Powles SB, Cornic G (1987) Aust J Plant Physiol 14: 539
149. Itoh K, Miyahara M (1984) Weed Res (Japan) 29: 301
150. Hickok CG, Schwarz OT (1989) Theor Appl Genet 77: 200
151. Powles SB (1987) Weed Res 26: 167
152. Mudge LC, Gossett BJ, Murphy TR (1984) Weed Sci 32: 591
153. Hess FD (1987) Rev Weed Sci 3: 193
154. Vaughn KC, Lehnen LP (1991) Weed Sci, in press
155. Vaughn KC (1986) Pestic Biochem Physiol 26: 66
156. Vaughn KC, Vaughan MA, Weeks DP (1987) Plant Physiol 83: 956
157. Vaughn KC, Vaughan MA (1990) Amer Chem Soc Ser 421: 364
158. Vaughn KC, Vaughan MA (1990) Long Ashton Symp Series, in press
159. Vaughn KC, Vaughan MA, Gossett BJ (1990) Weed Tech 4: 157
160. Morrison IN, Todd BG, Nowolsky KM (1989) Weed Tech 3: 544
161. Smeda RJ, Vaughn KC, Morrison IN (1991) Proc Weed Sci Soc Amer, 31: 47

162. Murphy TR, Gossett BJ, Toler JE (1986) Weed Sci 34: 704
163. Harrington KC (1987) Proc Aust Weeds Conf 126
164. Harrington KC (1990) Proc Ninth Aust Weeds Conf 201
165. Fedtke C (1982) Biochemistry and physiology of herbicide action, Springer, Berlin Heidelberg New York
166. Heim DR, Skomp JK, Tschabold EA, Larrinua IM (1990) Plant Physiol 93: 695
167. Delmer DP (1987) Plant Physiol 835: 109
168. Heim DR, Roberts JL, Pike PD, Larrinua FM (1989) Plant Physiol 90: 146
169. Heim DR, Roberts JL, Pike PD, Larrinua IM (1990) Plant Physiol 92: 858
170. Powles SB, Holtum JAM, Matthews JM, Liljegren DJ (1990) Amer Chem Soc Ser 421: 394
171. Holtum JAM, Matthews JM, Liljegren DJ, Hausler RE, Powles SB (1990) Proc Aust Weeds Conf 9: 224
172. Kemp MS, Moss SR, Thomas TH (1990) Amer Chem Soc Ser 421: 376
173. Powles SB, Holtum JAM (1990) Proc Aust Weeds Conf 9: 185

Subject Index

Previously Published Volumes in Chemistry of Plant Protection

Vol. 1: Sterol Biosynthesis, Inhibitors and Anti-Feeding Compounds

T. Kato,
Sterol-Biosynthesis in Fungi, a Target for Broad Sprectrum Fungicides

D. W. Krämer,
Chemistry of Sterol Biosynthesis Inhibiting Fungicides

K.-H. Kuck, H. Scheinpflug,
Biology of Sterol-Biosynthesis Inhibiting Fungicides

D. M. Norris,
Anti-Feeding Compounds

Vol. 2: Degradation of Pesticides, Desiccation and Defoliation, ACh-Receptors as Targets

P. R. Wallnöfer, G. Engelhardt,
Microbial Degradation of Pesticides

U. Schoknecht, D. Otto,
Enzymes Involved in the Metabolism of Organophosphorus, Carbamate and Pyrethroid Insecticides

R. Richter, D. Otto, H.-J. Mengs,
Insecticide Compounds, Acting on the Acetylcholine Receptor of the Insect Nervous System

H. Bergmann, D. Martin,
Chemical Manipulation of Desiccation and Defoliation and Essential Aspects for the Application and Development of New Chemical Compounds in the Future

Vol. 3: Pyrethroid Residues, Immunoassays for Low Molecular Weight Compounds

W. Blaß,
Residues of Pyrethroid Insecticides and Their Analytical Determination

E. W. Weiler,
Immunoassay for the Quantitation of Low Molecular Weight Compounds in Plants

Vol. 4

K. Naumann,
Synthetic Pyrethroid Insecticides: Structures and Properties

Vol. 5

K. Naumann,
Synthetic Pyrethroid Insecticides: Chemistry and Patents

Vol. 6: Controlled Release, Biochemical Effects of Pesticides, Inhibition of Plant Pathogenic Fungi

M. Bahadir, G. Pfister
Controlled Release Formulations of Pesticides

M. Eto
Biochemical Mechanisms of Insecticidal Activities

M. A. Q. Khan
Biochemical Effects of Pesticides on Mammals

P. Böger, G. Sandmann
Modern Herbicides Affecting Typical Plant Processes

H. Buchenauer
Physiological Reactions in the Inhibition of Plant Pathogenic Fungi

MIX
Papier aus verantwortungsvollen Quellen
Paper from responsible sources
FSC® C105338

If you have any concerns about our products,
you can contact us on
ProductSafety@springernature.com

In case Publisher is established outside the EU,
the EU authorized representative is:
Springer Nature Customer Service Center GmbH
Europaplatz 3, 69115 Heidelberg, Germany

Printed by Libri Plureos GmbH
in Hamburg, Germany